Photonics Rules of Thumb

Optical and Electro-Optical Engineering Series
Robert E. Fischer and Warren J. Smith, Series Editors

Published

Hecht
THE LASER GUIDEBOOK

Smith
MODERN OPTICAL ENGINEERING

Smith
MODERN LENS DESIGN

Waynant, Ediger
ELECTRO-OPTICS HANDBOOK

Wyatt
ELECTRO-OPTICAL SYSTEM DESIGN

Other Published Books of Interest

Optical Society of America
HANDBOOK OF OPTICS, SECOND EDITION, VOLUMES I, II

Keiser
OPTICAL FIBER COMMUNICATIONS

Syms, Cozens
OPTICAL GUIDED WAVES AND DEVICES

To order or receive additional information on these or any other McGraw-Hill titles, please call 1-800-822-8158 in the United States. In other countries, contact your local McGraw-Hill office.

WM16XXA

Photonics Rules of Thumb

Optics, Electro-Optics, Fiber Optics, and Lasers

John Lester Miller
Dr. Edward Friedman

McGraw-Hill

New York   San Francisco   Washington, D.C.   Auckland   Bogotá
Caracas   Lisbon   London   Madrid   Mexico City   Milan
Montreal   New Delhi   San Juan   Singapore
Sydney   Tokyo   Toronto

Library of Congress Cataloging-in-Publication Data

Miller, John Lester, date
   Photonics rules of thumb : optics, electro-optics, fiber optics,
and lasers   /   John Lester Miller, Edward Friedman.
      p.   cm. — (Optical and electro-optical engineering series)
   Includes index.
   ISBN 0-07-044329-7 (hc)
   1. Photonics. I. Friedman, Edward. II. Title. III. Series.
TA1520.M55 1996   96-10905
621.36—dc20   CIP

McGraw-Hill

A Division of The McGraw·Hill Companies

ISBN 0-07-044329-7

*The sponsoring editor for this book was Stephen S. Chapman and the production
supervisor was Pamela A. Pelton. It was set in ITC New Baskerville by J. K.
Eckert & Company, Inc.*

Printed and bound by R. R. Donnelley & Sons Company.

This book is printed on recycled, acid-free paper containing a
minimum of 50 percent recycled de-inked fiber.

Contents

Preface

John Lester Miller and Dr. Ed Friedman arrived at the same place by very different paths. John spent some of his career in astronomy before joining an aerospace company to work on infrared sensors for space surveillance. Ed spent much of his career working on remote sensing technologies applied to the Earth, its atmosphere, and oceans. The two met in Denver in 1985, both working for a major government contractor on exotic electro-optical systems.

Those were halcyon days, with money flowing like water and contractors winning billions of dollars for some concepts that were overly optimistic or barely possible at best. In the center of the whole fray were bureaucrats and military officers who were demanding that we design systems capable of the impossible. The authors saw many requirements and goals being levied on our systems that were far from realistic and often resulted from confusing interpretations of the capabilities of optical and electro-optical systems

The authors found a common ground when they discovered that many co-workers, in an attempt to outdo the competition, were promising to perform sensor demonstrations that violated many rules of engineering, if not physics. On one dubious billion-dollar program, after some consistent exposure to neophytes proposing all sorts of undoable things, they decided to try to educate everyone by creating a half-serious, half-humorous posting for the local bulletin board called "Dr. Photon's Rules of Thumb." Its content was a list of basic rules that apply when optics or electro-optics are being used.

The list consisted of the worst of the erroneous ideas that nontechnical people had proposed. The goal was to eliminate many of the bad ideas and try to instill at least a little scientific foundation to the efforts of the team. Although the list of simple rules embarrassed a few of the misinformed, it was generally met with enthusiasm, and Messrs. Miller and Friedman found copies of it not only in their project but all across the company, and even external to the company.

So, what is the value of a collection of rules? The answer lies in the way optical and electro-optical designs are carried out these days. No one works alone. Commonly, a design study starts with a team meeting that includes representatives of all of the engineering disciplines that will work on the project. These days, the "concurrent engineering" concept is all the rage and implies that everyone who will ever work on the design from beginning to end is present at key meetings.

Next, usually a system engineer who has studied the customer requirements lays out a concept for the design team to improve and provide details. Wouldn't it be nice if each person came to these meetings prepared to detect fundamentally flawed approaches and identify and eliminate them early in the design effort? For instance, the authors encountered a number of situations in which a tracker's target signature was known to be a strong function of aspect angle. Ignoring this, the system engineer had chosen an engagement geometry that would have required a sensor with an impractical aperture.

Therefore, the intent of this book is to help the optics and E-O parts of those teams make quick assessments, using no more than a calculator, so that they don't have to try to support a fundamentally flawed concept. The book is also useful for managers, marketeers, and other semitechnical folks who are new to the electro-optical industry (or are on its periphery) to develop a feel for the difference between the chimerical and the real. Students may find the same type of quick-calculation approach valuable, particularly in the case of oral exams where the professor is pressuring the student to do a complex problem quickly. Using these assembled rules, you can keep your wits about you and provide an immediate and nearly correct answer, which usually will save the day. After the day is saved, you should go back to the question and perform a rigorous analysis. These rules are useful for quick sanity checks and basic relationships. Being familiar with the rules allows one to rapidly pinpoint trouble areas or ask probing questions in meetings. They aid in thinking on your feet and in developing a sense of what will work and what won't. But they are not, and never will be, the last word. It is fully recognized that errors may still be present, and for that the authors apologize in advance to readers and those from whom the material was derived.

The book begins with an introduction, which discusses what a rule is, how to use it, and when and why they are appropriate. The bulk of this book consists of 234 rules, divided into 17 chapters. Each chapter begins with a short background and history of the general subject matter to set the stage and provide a foundation. The rules follow. If there are additional references for the rule, they can be found immediately following the rule. Because many rules apply to more than one chapter, a comprehensive index and detailed table of contents is included. The authors apologize for any confusion you may have in finding a given rule, but it was necessary to put them somewhere and, quite honestly, it was often an arbitrary choice between one or more chapters. Students and those new to the field will find the glossary useful. Here you will find definitions of jargon, common acronyms, abbreviations, and a lexicon of confusing and ambiguous terms. The book closes with a bibliography of what we feel to be important current textbooks in the field.

The quest for rules led to key papers, hallmark books, colleagues, and into the heart of darkness of past experience. Some of these rules are common folklore in the industry, whereas others have been developed by the authors. The book has been assembled from a wide variety of sources including experts, out-of-print books, foreign books, and technical papers. It is by no means comprehensive, but it does provide some key rules in a wide range of fields. The original list included nearly 500 such rules. Rule selection was based on the authors' perceptions of each rule's practical usefulness to electro-optical professionals in the late 1990s. The downselect was accomplished by examining every rule for truthfulness, practicality, range of applicability, ease of understanding, and, frankly, ones that they just liked. As such, this is an eclectic assortment that will be more useful to some than others, and more useful on some days than others. Many rules have been combined into a single rule. Therefore, some rules have several nearly identical equations or concepts describing the same concept.

This collection of rules and concepts represents an incomplete, idiosyncratic, and eclectic toolbox. The rules, like tools, are neither good nor bad; they can be used to facilitate the maturing of whimsical concepts to mature hardware or to immediately identify a technological path worth pursuing. Conversely, they can also be used to obfuscate the truth and, if improperly applied, to derive incorrect answers. The job was to refine complex ideas to simplified concepts and present these rules to you with appropriate cautions. However, it is your responsibility to use them correctly. Remember, it is a poor workman who blames his tools.

Acknowledgments

The authors have many to whom they owe a debt; virtually every rule contained in this book is due to the specific contribution or the inspiration of someone in the field of electro-optics. Without their efforts to develop the technology and its many applications, this book would neither exist nor have any value. We also acknowledge the contribution made by the many reviewers who helped remove errors and improve the content: Max Amon, W.M. Bloomquist, Joe Calbretta, Gene Dryden, Mike Elias, Phil Ely, Corinne Foster, Doug Franzen, Dr. Grant Gerhart, Dr. Jim Harrington, Jerry Holst, Dr. Walt Kailey, Dr. J. Richard Kerr, Dr. Don Malakoff, Dr. Robert Martin, Judy McFadden, Grant Milbouer, Bruce Mitchell, Doug Murphy, Dr. Ray Radebaugh, Tom Roberts, Greg Ronan, Ken Sarkady, Wade Scherer, Dr. George Spencer, and Dr. Gerry Steiner. Separately, we also thank all of those who suggested rules and provided permissions. Finally, the authors recognize the role of our families (especially our beloved wives, Judith Friedman and Corinne Foster) for tolerating the long periods of loneliness and being barred from our computers while this book was created.

Introduction

"Few formulas are so absolute as not to bend before the blast of extraordinary circumstances."
BENJAMIN NATHAN CARDOZO

The evolution of the electro-optical (EO) sciences parallels, and feeds from, developments in a number of somewhat unrelated fields, including astronomy, satellite and remote sensing technology, materials science, electronics, optical communications, military research, and many others. The common thread of all of this effort, which really came into focus in the early 1950s, is that scientists and engineers have been able to apply the highly successful electronic technologies with the more ancient concepts and methods of optics and electromagnetic wave propagation. The merging of these fields has provided an unprecedented capability for instruments to "see" targets and communicate with them in a wide range of wavelengths for the benefit of science, defense, and (more recently) consumers. Major departments at universities are now devoted to producing new graduates with specialties in this field. There is no end in sight for the advancement of these technologies, especially with the continued development of electronics and computing as increasingly integral parts of EO instrumentation.

One of the disturbing trends in this technology is the constant narrowing of the role of engineers. As the technology matures, it becomes harder for anyone working in an area of the EO discipline to understand all of what is being done in EO sciences and engineering. This book has been assembled to make a first, small step on a wide range of topics accessible to anyone working in EO. There is no intent to compete with the *IR/EO Handbook, The Handbook of Optics,* or the many journals devoted to EO fields, all of which provide considerable detail in every area. On the other hand, it is intended to allow any EO engineer, whatever his specialty, to make first guesses in a wide range of topics that might be encountered in system design or fabrication.

The motivation for the book was to provide a forum for engineers working in the electro-optical fields to record those ideas that allow them to quickly assess if a design idea will work, what it will cost, and how risky it is. All of us do this, but we usually don't organize our own set of rules and make them public. To assist us in this endeavor, we have solicited the cooperation of as many experts as would agree to help. Their input gives us a wide variety of experience from many different technical points of view.

Alas, technology advances, and all of us wonder how we can possibly keep up. Hopefully, this book will not only provide some specific ideas related to electro-optic technology, but will also suggest some ways of thinking about things that will lead to a whole new generation of such rules and ideas.

As our search to find rules began, we realized that not everyone has the same thing in mind when providing one. To qualify for our definition of a *rule of thumb*, a rule should be

- easy to implement
- able to provide roughly the correct answer
- easy to remember
- simple to express
- useful to a practitioner

In documenting the rules, we found it valuable to create a standard form and stick to it as closely as possible. There are some violations on the theme in the book but, on the whole, rules are presented in the following standard format.

Name of The Rule

A descriptive name for the rule, which will help the reader quickly locate a rule in this book.

Subject

A brief list of the topical areas for which the rule might apply

The Rule

A succinct statement of the rule, followed by an equation if appropriate.

Basis for the Rule

This usually ended up as a brief justification of the rule and is intended to provide details enough to let the interested user follow up with additional reading. Sometimes, where appropriate several sentences explaining the history of the scientific derivation are included here.

Cautions and Useful Range of the Rule

Any limitations, caveats, assumptions or cautions in applying the rule are listed as well as the useful range of the rule.

Usefulness of Rule

A list of situations or conditions where the rule can be useful.

Notes and Explanation

Here is presented any ancillary information and details not covered in the above.

Example (if appropriate)

When useful, an example is provided.

The rules in this book are intended to be correct in principle and to give the right answer to an approximation. Some are more accurate than others. Readers who have to get the answer more precisely can sometimes refer to the given source of the rule, which usually contains some more detailed analyses that give an even better answer.

Rules based on the current state of art will be useful in the future only to demonstrate how hard a Twentieth Century infrared engineer had to work in these current Dark Ages. These will all become less useful as time marches on, but they are valid now and for the rest of this century. However, even today, there may arise odd situations when a particular rule will be invalid. When this happens, a detailed understanding from management to technician must exist as to why the state of the art is being beaten. It isn't impossible to beat the state of the art—only unlikely (unless you are trying).

The book is organized into 17 chapters covering most of the important topics that EO engineers encounter regularly. The number of rules per chapter ranges from about 5 to nearly 40. Clearly, this does not cover all of the disciplines related to this diverse field. But, it is a start.

1

Astronomy

This chapter contains a selection of rules specifically involving the intersection of the disciplines of astronomy and electro-optics (EO). Sensors frequently look upward, so astronomical objects frequently define the background for many systems. Moreover, many sensors are specifically designed to detect heavenly bodies, so astronomical relationships define the targets for many sensors.

Over the past few hundred years, astronomy has driven photonics, and photonics has enabled modern astronomy. The disciplines have been as interwoven as DNA. Frequently, key discoveries in astronomy are impossible until photonic technology develops to a level that permits them. Conversely, photonic development often has been funded and refined by the astronomical sciences as well as the military. Military interest in detecting and tracking satellites has been an important source of new technology that furthers the application of electro-optics in astronomy. Some say that one of the most important contributions of the Strategic Defense Initiative (SDI) was the advancement of certain photonic technologies that are currently benefiting astronomers. Some of these include adaptive optics, synthetic reference stars, large and sensitive focal planes, advanced materials for astronomical telescopes, new methods of image stabilization, and advanced computers and algorithms for interpreting images distorted by atmospheric effects. Thus, the importance of astronomy to the development of electro-optics cannot be overestimated.

The new millennium will include a host of new-technology telescopes that may surpass space-based observation capabilities (except in the spectral regions where the atmosphere strongly absorbs or scatters). The two Keck 10-meter telescopes represent an amazing electro-optical engineering achievement. By employing segmented mirrors, lightweight mirrors, lightweight structure, and by adjusting the mirrors in real time, we have been able to discard many of the past notions and operating paradigms about telescopes. Soon, the Kecks will be eclipsed by larger and more powerful phased arrays of optical telescopes, all using new photonic technology that was not even dreamed of 20 years ago. This new emphasis on novel technology applied to Earth-based telescopes represents a major addition to the astronomical community's toolbox, and a shift in the electro-optical and astronomical communities' perceptions. Most of the advancements that will be employed were developed by military research and funding, not specifically for astronomy. In the near future, these high-technology telescopes, coupled with advanced precision instruments, will provide astronomers with new tools to make new and wondrous discoveries.

For the reader interested in more details, there is a myriad of observational astronomy books, but few deal directly with observational astronomy utilizing electro-optics. For specific EO discussions, an uncoordinated smattering can be found throughout the *EO/IR Handbook* and Miller's *Principles of Infrared Technology*. Additionally, Schroeder's *Astronomical Optics* addresses many principles in detail. Do not forget to check the journals, as they seem to have more material relating to EO astronomy than do books. The Society of Photo-Optical Instrumentation Engineers (SPIE) regularly has conference sessions on astronomical optics, instruments, large telescopes and such, and the Astronomical Society has several west-coast conferences featuring many EO-related papers. Additionally, there are many good articles and papers appearing in *Sky & Telescope, Infrared Physics,* and the venerable *Astrophysical Journal.* Finally, do not overlook World Wide Web sites on the Internet; much good information is made available by several observatories.

BLACKBODY TEMPERATURE OF THE SUN

Subject: Astronomy, radiometry, phenomenology

The Rule

Consider the Sun a 6000 K blackbody.

Basis for the Rule

This rule is based on empirical observations and simple stellar models.

Cautions and Useful Range of the Rule

This rule represents an estimate only; the sun may appear as a blackbody with a temperature anywhere from about 5750 and 6100 K, depending upon wavelength.

The Sun (and all stars) do have some absorption and emission lines, so using a blackbody approximation is valid only for broad bandpasses.

Moreover, the atmosphere strongly absorbs and transmits in various windows. The above rule is derived from a general curve fit for wide bandpasses, disregarding atmospheric absorption.

Usefulness of the Rule

This is useful for estimating the flux from the sun using Planck's equation.

Notes and Explanation

The sun is a complex system. It is a main-sequence star (G2) of middle age. It has several layers of different temperatures and different energy-producing mechanisms. Almost all the light we see is emitted from a thin layer at the surface. Temperatures can reach over 10 million degrees in the center. However at the surface the temperature is usually something under 6000 K, with the best fit to a blackbody curve usually at 5770 K.

A star's blackbody temperature based on spectral type is (to a first order) approximately: B:27000, A: 9900, F:7000, G:5900, K:5200, and M 3800 K.

Incidentally, the moon approximates a blackbody with a 400 K temperature.

ATMOSPHERIC "SEEING"

Subject: Astronomy, atmospherics, systems

The Rule

Typically, the best seeing is just under 1 arcsecond, or about 4 μrad.

Basis for the Rule

This is based on empirical astronomical observations from too many cold nights on mountain tops.

Cautions and Useful Range of the Rule

"Seeing" tends to improve with wavelength (to something like the 6/5th power) That is, the seeing angle gets smaller (better) as the wavelength increases. Also see the various rules in the Chapter 2, Atmospherics.

This does not account for any atmospheric stabilization techniques.

The rule is limited to Earth-based observations with especially clear weather.

The world's record for "seeing" is somewhat better. Although very rare, seeing may approach 0.1 to 0.2 arcseconds (or around 0.5 to 1 μrad) with ideal conditions.

This does not apply to speckle interferometry, which is not limited by such conditions.

Usefulness of the Rule

This is useful for determining the actual resolution likely to be obtained.

Notes and Explanation

Frequently, resolution drivers are not diffraction limit criteria, but are of being "atmospheric-seeing-limited." This "seeing" tends to improve with increasing altitude and wavelength. It can be much worse if the atmosphere is cloudy or turbulent.

When an object is viewed through the atmosphere, the atmosphere degrades the image by convection, turbulence, and varying index of refraction. This induces a variety of effects, including wavefront tilt, scintillation, and blurring. Typical "seeing" obtainable on good nights, at high altitude observatories, is approximately one arcsecond. This limit is a function of the atmosphere; the aperture size is often of no consequence. In fact, the common amateur telescope is well matched to the limits imposed by the atmosphere at low altitude. Large, professional telescopes are advantageous because they can collect more light and employ advanced techniques such as adaptive optics to clean up the images, and because they are located at premier locations (e.g., Mauna Kea).

PHOTON RATE AT A TELESCOPE APERTURE

Subject: Astronomy, stellar irradiance, backgrounds, phenomenology

The Rule

The photon rate at a telescope's aperture from a star of magnitude m is:[1]

$$S = Nt\frac{\pi}{4}(1 - \varepsilon^2)\, D^2 \Delta\lambda\, 10^{-0.4\,m}$$

where S = photon flux in photons/second
N = irradiance of a magnitude zero star [this is $\approx 1 \times 10^7$ photons/$(\mathrm{cm}^2 \sec \mu)$]
D = diameter of the telescope in cm
t = unitless transmittance of the atmosphere and optics
$\Delta\lambda$ = bandpass of interest in nm
m = visual magnitude of the star
ε = obscuration ratio

Basis for the Rule

This rule is founded in basic radiometry, geometry, optics, and the definition of stellar magnitude.

Cautions and Useful Range of the Rule

This was developed for AO-class stars. It is valid for narrow visible bandpasses. (Use with caution elsewhere.)

Usefulness of the Rule

It is useful for quick estimates of the expected flux at an aperture from a source and to determine inputs for signal-to-noise ratio calculations or to determine the limiting magnitude of the telescope.

Notes and Explanation

The above rule allows the approximate calculation of the number of photons per second at the telescope aperture. Additionally, Kailey[*] gives us some handy approximations: a difference of one magnitude results in a difference of about 2.5 in spectral irradiance, a difference of 5 magnitudes is a factor of 100 difference in spectral irradiance, and a small magnitude difference is equivalent to an equal percentage difference in brightness (10.01 magnitudes is ≈ 1 percent dimmer than 10.00 magnitudes).

Most on-axis reflecting telescopes have circular obscurations in the center of the aperture. Therefore, Schroeder suggests that $[\pi(1 - \varepsilon^2)]/4$ for a Cassegrain telescope, so the equation simplifies to $S = 0.7Nt\, D^2 \Delta\lambda\, 10^{-0.4m}$.

Reference

1. D. Schroeder. 1987. *Astronomical Optics.* Orlando, FL: Academic Press, 319.

---

[*]Information supplied by Dr. Walt Kailey, 1995.

NUMBER OF STARS AS A FUNCTION OF WAVELENGTH

Subject: Astronomy, backgrounds, phenomenology

The Rule

At visible wavelengths or beyond, for a given sensitivity, the longer the wavelength, the fewer stars that you can sense. The fall off in the number of stars approximates:

$$\#S_{\lambda_2} = \#S_{\lambda_1} 10^{-0.4R}$$

$\#S_{\lambda_2}$ where = number of stars at wavelength λ_2 (λ_2 larger than λ_1) at a given irradiance

R = ratio of one wavelength to another, λ_2/λ_1

$\#S_{\lambda_1}$ = number of stars at wavelength λ_1

Basis for the Rule

This rule is based on curve fitting empirical data. Curves supporting this can be found in Seyrafi's *Electro-Optical Systems Analysis* and Hudson's *Infrared Systems Engineering.*

Cautions and Useful Range of the Rule

This is useful for separate narrow bands, from about 0.7 to 15 μm, and irradiance levels on the order of 10^{-13} W/cm^2/μm. Generally, this provides accuracy to within a factor of two.

Usefulness of the Rule

This is useful for scaling the population of stars from one wavelength to another.

Notes and Explanation

The authors curve fitted data to derive the above relationship. This rule is a less accurate and more general version of the same concept as the "Number of Stars above a Given Irradiance" rule (page 8). However, that rule emphasizes the irradiance distribution, while this rule emphasizes the wavelength distribution. Most stars are like our sun and radiate most of their energy in what we call the visible part of the spectrum. As wavelength increases, there are fewer stars because the Planck function is dropping for the stars that peak in the visible, and fewer stars have peak radiant output at longer wavelengths.

NUMBER OF STARS AS A FUNCTION OF VISUAL MAGNITUDE

Subject: Astronomy, phenomenology, backgrounds

The Rule

The number of stars brighter than a particular visual magnitude can be estimated from the following simple relationships:

1. Number of visual stars brighter than a given magnitude $\approx 2 \times e^{\text{magnitude}}$

2. Another curve fit provides the following relationship:

$$\#S = 11.8 \times 10^{(0.4204 \, M_v)}$$

where $\#S$ = approximate number of stars
M_v = visual magnitude

Basis for the Rule

These equations were based on curve fitting to empirical data.

Cautions and Useful Range of the Rule

These are accurate to within a factor of 3 between magnitude 0 and 20 in a visual bandpass.

Usefulness of the Rule

This is useful for estimating stars detectable for background calculations, star surveys, attitude control systems, or clutter rejection purposes.

Notes and Explanation

The authors have curve fitted the number of stars versus their magnitude and developed the above relationships. The first is easier to remember but diverges from reality at dimmer magnitudes. The second equation provides a good match (better than ±30 percent) for most magnitudes. However, the second equation tends to underpredict the number of stars between magnitudes 13 and 15 and overpredict the number of stars between magnitudes 16 and 20.

NUMBER OF STARS ABOVE A GIVEN IRRADIANCE

Subject: Astronomy, backgrounds, infrared observation of stars

The Rules

1. The total number of stars at or above an irradiance of 10^{-13} W/cm^2/µm, is approximately $1400 \times 10^{(-0.626\lambda)}$, where λ is the wavelength in microns.

2. The number of stars at or above 10^{-14} W/cm^2/µm, is approximately $4300 \times 10^{(-0.415\lambda)}$, where λ is the wavelength in microns.

3. The number of stars at or above a radiance of 10^{-15} W/cm^2/µm, is approximately $21,000 \times 10^{(-0.266\lambda)}$, where λ is the wavelength in microns.

Basis for the Rules

These equations were based on curve fitting crude empirical data.

Cautions and Useful Range of the Rules

These simple equations seem to track the real data within a factor of two (or three) within the wavelength bounds. The curve for 10^{-15} W/cm^2/µm tends to underpredict below 4 µm. The 10^{-13} W/cm^2/µm curve was based on data from 1 to 4 µm, the 10^{-14} equation on wavelengths of 2 to 8 µm, and the 10^{-15} on wavelengths from 2 to 10 µm.

Usefulness of the Rules

This provides guesstimates for sensing of stars which can be used for IR star tracker sensitivity requirements, and ATH background estimations.

Notes and Explanation

The authors curve fitted some reasonably crude data to derive the above relationships. This rule highlights two phenomena. First, the longer the wavelength (beyond visual) that you observe, the fewer stars you will detect at a given magnitude or irradiance. Second, increased instrument sensitivity provides an increased number of stars detected.

As one observes in longer and longer wavelengths, there are fewer stars to observe at a given brightness. This phenomenon stems from stellar evolution and Planck's theory. Most stars fall into what astronomer's call "the main sequence" and have their peak output between 0.4 and 0.8 µm. From Planck's equations, we also note that the longer the wavelength, the less output (brightness) a star has for typical stellar temperatures—on the tail of the Planck function.

NUMBER OF INFRARED SOURCES PER SQUARE DEGREE

Subject: Astronomy, phenomenology, backgrounds

The Rule

$$N_s \approx \log N_\lambda [s(b)] = \log [A(b, L)] + B(b, L) \log \{E_{12}[\lambda, s(b)]\}$$

where N_s = number of infrared sources per square degree
$\{E_{12}[\lambda, s(b)]\}$ = equivalent spectral irradiance at 12 µm producing $N_\lambda\{s(b)\}$
    sources per square degree, Jansky
b = galactic latitude, $0° \le b \le 90°$
L = galactic longitude in degrees, $0° \le L \le 180°$

and

$$\log [A(b,L)] = 0.000488L - 0.78 + [(0.000061L^2 - 0.0208L + 3.0214)/ \\ (1 + \{b/12\}^{1.4})]$$
$$B(b,L) = (-0.00978L + 0.88)\{1.0 - \exp(-b/(8 - 0.05L))\} + (0.00978L - 1.8)$$

Basis for the Rule

This is based on empirical observations from IRAS sources.

Cautions and Useful Range of the Rule

For $L > 90°$, the $B = -0.92$

Usefulness of the Rule

This is useful for estimating the sources in the sky.

Notes and Explanation

The IR sky is rich with astronomical sources. This complicated rule provides an excellent match to the distribution of sources found in nature. Note that it is the function of the spectral index portion of the equation to extend the model to other wavelengths. To do so, the spectral energy distribution of the mean ensemble of sources must be known.

The unit "Jansky" refers to flux density units. To convert from Jansky to spectral irradiance, use the expression

$$E_\lambda \left(\frac{w}{cm^2\mu}\right) = 3 \times 10^{-16} \lambda^{-2} \bullet \text{Jansky}$$

where λ = wavelength in microns. Therefore,

$$1.5 \text{ Jansky} = 3.125 \times 10^{-18} \frac{w}{cm^2\mu} \quad \text{at } 12\ \mu$$

Reference

1. D. D. Kryskowski and G. Suits. 1993. Natural sources. In *Sources of Radiation*, vol. 1, ed. G. Zissis, of *The Infrared and Electro Optical Systems Handbook,* executive ed. J. Accetta and D. Shumaker. Ann Arbor, MI: ERIM, and Bellingham, WA: SPIE, 179–180.

Atmospherics

It is hard to imagine a subject more complex, and yet more useful, than the propagation of light in the atmosphere. Due to its importance in a wide variety of human enterprises, there has been considerable attention paid to this topic for several centuries. Initially, the effort was dedicated to learning how the apparent size and shape of distant objects depend on the properties of the atmosphere. The eventual understanding of quantum mechanics led to a formal understanding of the absorption spectra of significant atmospheric species and their variation with altitude. Computer models that include virtually all that is known about the absorption and scattering properties of atmospheric constituents have been assembled and can provide very complete descriptions of transmission as a function of wavelength with a spectral resolution of about 1 cm^{-1}. This is equivalent to a wavelength resolution of 0.1 nm at a wavelength of 1 micron.

In addition to gradually refining our understanding of atmospheric absorption by considering the combined effect of the constituents, we also have developed a rather complete and elaborate theory of scattering in the atmosphere. The modern model of the scattering of the atmosphere owes its roots to the efforts of Mie and Rayleigh. Their results have been extended greatly by the use of computer modeling, particularly in the field of multiple scattering and Monte Carlo methods. The result is that, for suspended particulates of known optical properties, reliable estimates of scattering properties for both plane and spherical waves can be obtained for conditions in which the optical thickness is not too large.

Two technologies have been in the background in all of these theoretical developments: spectroscopy and electro-optical technology. Spectroscopes, which are essential instruments in the measurement of the spectral transmission of the atmosphere, are EO systems, relying on improvements in detectors, optics, control mechanisms, and many of the other topics addressed in this book. As these technologies have matured, continuous improvements have been seen in the ability to measure properties of the atmosphere and to turn those results into useful applications. The applications range from measuring the expected optical properties being considered for astronomical telescope location to determining the amount of sunlight that enters the atmosphere and is subsequently scattered back into space. Laser technology has also been a key factor in new measurements of atmospheric behavior, in both scattering and absorption phenomena. Tunable lasers with very narrow line widths have been employed to verify atmospheric models and have provided the ability characterize not only the "clean" atmosphere but the properties of atmospheres containing aerosols and industrial air pollution. Laser radar is regularly used to characterize the vertical distribution of scattering material, cloud density, and other features of the environment, weather, and climate.

New advances in EO technologies have also allowed new insight into the radiation transfer into, out of, and within the atmosphere. Satellite-based remote sensors have been measuring the radiation budget of the earth in an attempt to define its impact on the climatic trends. There are many other examples of space-based EO sensors that concentrate on measuring properties of the atmosphere, including the concentration of trace constituents in the stratosphere, ozone concentrations over the poles, and so on.

Recently, at the urging of the military, measurements and improved theory have led to the development of methods for estimating and removing clear-air turbulence effects. These effects have an influence on the ability of a telescope to form an image or participate in optical communications. New advancements in measuring the wavefront errors resulting from turbulence are included in adaptive optics. This technology is able to remove, up to a point, adverse atmospheric effects, which leads to telescope images that parallel those that would occur in transmission through a vacuum.

We can expect continual improvement in our understanding of the atmosphere and the way that it interacts with light propagating in it. All of these improvements in theory, supported by advancements in instrumentation quality, will result in even more capable EO systems and allow them to reduce the perturbing effects of their operating environments.

The rules in this chapter provide a sample of some of the concepts that have evolved from our current understanding of the atmosphere, with particular emphasis on the propagation of laser radiation. This is appropriate because many of the systems that involve long-distance propagation of radiation for imaging or communication rely on laser systems.

The interested reader can find technical articles in *Applied Optics* and similar technical journals. At the same time, magazines such as *Sky & Telescope* and *Popular Science* occasionally include information on the way astronomers are using new technologies to cope with the effects of the atmosphere. A few new books have come out that deal specifically with imaging through the atmosphere. The Society of Photo-Optical Instrumentation Engineers (SPIE) is a good source for these texts.

ATMOSPHERIC ATTENUATION, OR BEER'S LAW

Subject: Atmospherics, lasers, system performance

The Rule

The attenuation of a beam of light, traversing an attenuating medium often can be estimated by the simple form:

$$\text{Transmission} = e^{-\alpha z}$$

where α = attenuation coefficient
$\qquad z$ = path length

This common form is called *Beer's law* and is useful in describing the attenuation of a beam in atmospheric and water environments, and in optical materials. Since both absorption and scattering will remove energy from the beam, α is usually expressed as:

$$\alpha = a + \gamma$$

where a = absorption per unit length
$\qquad \gamma$ = scattering per the same unit length

Basis for the Rule

The equations result directly from Beer's law. This law is common with transmission problems associated with most media. The idea is that scattering and absorption remove light from a collimated beam at a rate proportional to the thickness of the path and the amount of scattering or absorbing material that is present.

The rule is derived from the fact that the fractional amount of radiation removed from the beam is independent of the beam intensity. This leads to a differential equation of the form (dz/z) = constant. The solution of this simple equation is of exponential form. The numerical values in the equations are derived from field measurements.

Cautions and Useful Range of the Rule

Be aware that the application of this universal rule assumes that the radiation in question is in the form of a beam. For example, downwelling sunlight in the atmosphere or ocean is described by a different attenuation coefficient that must take into account the fact that the scattered light is not removed from the system but can still contribute to the overall radiation. In the beam case, scattering removes light from the beam in an explicit way.

Make sure all units agree, and be sure to use the correct bandpass when determining the scattering coefficient.

If changes of altitude are involved, be sure to note that α depends on altitude in the atmosphere.

Rarely can one number be used to represent total in-band transmission. Such simplifications are inevitable when a simple set of equations is used to model a complex system such as the atmosphere.

Usefulness of the Rule

This rule provides a quick estimate of atmospheric transmission.

Notes and Explanation

The presence of an exponential attenuation term in transmission through the atmosphere is no surprise, given that this mathematical form appears for a wide variety of media (including the bulk absorption of optical materials such as glass). The ability to estimate attenuation in the presence of rain is convenient for many applications. The link in the atmosphere is subject to considerable limitation due to small particles, even if the rainfall rate is low.

The value of γ due to rain can be estimated in the following way:

$$\gamma \approx 3.92/V$$

where V = visual range (usually defined as contrast between target and background of 2 percent). For cases of Mie scattering, γ varies as λ^{-4}.

Hudson[1] also shows that the effect of rain on visual range and scattering coefficient can be estimated from

$$s = (1.25 \times 10^{-6}) \frac{Z}{r^3}$$

where Z = rainfall rate in cm/s
      r = radius of the drop in cm

Alternatively, Burle[2] gives the scattering coefficient of rainfall as:

$$\gamma = 0.248 f^{0.67}$$

where f = rainfall rate in mm/hr

References

1. R. Hudson. 1969. *Infrared Systems Engineering.* New York: John Wiley & Sons, 161–165.
2. Burle Industries. 1974. *Burle Electro-Optics Handbook.* Lancaster, PA: Burle Industries, 87.

ISOPLANATIC ANGLE

Subject: Atmosphere

The Rule

The effects of atmospheric turbulence can be summarized in a single parameter that defines the angular extent over which its effect is constant. This parameter is called the *isoplanatic angle*. It is defined below.

Basis for the Rule

This rule is a simplifying characterization of the properties of the atmosphere.

Cautions and Useful Range of the Rule

The complexities of the properties of the atmosphere have been simplified in a number of ways to produce this type of characterization. The most profound assumption derives from the use of the Kolmogorov description of the statistical behavior of fluid turbulence. Extensive research in this field has confirmed most of the results predicted by his theory, but its application is not universal.

Usefulness of the Rule

The definition of isoplanatism is of particular importance to system designers who intend to use stars and other astronomical objects to correct for the degradations induced by the atmosphere. The angle above provides a measure of how far apart, in angle, an object to be viewed can be from the reference star.

Notes and Explanation

The isoplanatic angle is defined as the range of angle of rays entering a telescope over which the mean square wavefront distortion varies by 1 radian. Expressed another way, it is the angle between two beams of light such that the phase difference is small enough that adaptive optics in the receiving telescope can correct the problem. It relies on Fried's parameter r_o, which is described elsewhere in this chapter.

The angle is defined as

$$\theta_o = \left(2.91\ k^2 \int_0^L C_n^2(z)\, z^{5/3}\, dz \right)^{-3/5}$$

where L = distance to the layer of turbulence
$\quad k$ = propagation constant = $(2\pi/\lambda)$
$\quad C_n^2$ = index of refraction structure constant

When C_n^2 is constant, we can simplify the expression by using Fried's parameter, which equals

$$(0.423 k^2 C_n^2 L)^{-3/5}$$

and we get

$$\theta_o \approx 0.6 \frac{r_o}{L}$$

Typically, the isoplanatic angle is in excess of 1 arcsecond but is smaller than 5 arcseconds. Because atmospheric conditions determine the value of the isoplanatic angle, it varies from place to place and time to time. It is interesting to invert the above equation to see the effective distance through which a ground-based optical telescope views a star. If r_0 is about 15 cm and θ_0 is about 1 arcsecond (5 μrad), we conclude that L is about 18 km, which is the altitude at which C_n^2 approaches zero.

Source

R. Tyson. 1991. *Principles of Adaptive Optics.* San Diego, CA: Academic Press, 83.

C_n^2 ESTIMATES

Subject: Atmospherics, lasers, systems

The Rule

The index of refraction structure constant C_n^2 is a measure of the index of refraction variation induced by small-scale variations in the temperature of the atmosphere. No other parameter is so common as C_n^2 in defining the impact of the atmosphere on the propagation of light. There are several quick ways to estimate C_n^2. The easiest is that is varies with altitude in the following way:[1]

$$C_n^2(h) = \frac{1.5 \times 10^{-13}}{h \, (\text{in meters})} \quad \text{for } h < 20 \text{ km}$$

and

$$C_n^2(h) = 0 \text{ for altitudes above 20 km}$$

Note that $C_n^2(h)$ has the rather odd units of $m^{-2/3}$.

Basis for the Rule

$C_n^2(h)$, in which we have explicitly shown that the function depends on altitude, is critical in determining a number of key propagation issues (such as laser beam expansion, visibility, and adaptive optics system performance), and defines the performance impact that the atmosphere has on ground-based astronomical telescopes.

Cautions and Useful Range of the Rule

Each of the estimates shown below is just that—an estimate. However, for the modeling of systems or the sizing of the optical components to be used in a communications or illumination instruments, these approximations are quite adequate. This field of study is particularly rich, having benefited from work by both the scientific and military communities. Any attempt to provide really accurate approximations to the real behavior of the atmosphere is beyond the scope of this type of book, but the subject is covered frequently in the astronomical literature, particularly in conferences and papers that deal with the design of ground-based telescopes.

Usefulness of the Rule

Beam propagation estimates rely on knowledge of C_n^2. Many of those estimation methods appear in this chapter. Although there are a number of estimates of C_n^2 that are widely used, most will provide adequate results for system engineers trying to determine the impact of turbulence on the intensity of a beam as well as other features of beam propagation.

Notes and Explanation

The most widely used analytic expression for $C_n^2(h)$ is the so-called *Hufnagel-Valley (HV) 5/7 model*. It is so-called because the profile of C_n^2 results in a Fried parameter (see the rule "Fried parameter") of 5 cm and an isoplanatic angle (see the rule "isoplanatic angle") of 7 μrad for a wavelength of 0.5 microns. Beland[2] expresses the Hufnagel-Valley (HV) 5/7 model as

$$8.2 \times 10^{-26} \left(\frac{h}{1000} \right)^{10} W^2 \exp\left(-\frac{h}{1000} \right) + 2.7 \times 10^{-16} e^{-h/1500} + 1.7 \times 10^{-14} e^{-h/100}$$

where h = height in meters
W = wind correlating factor, which is selected as 21 for the HV 5/7 model

Note that the second reference has an error in the multiplier in the last term. That error has been corrected in the material presented above.

In many cases, C_n^2 value can be crudely approximated as simply 1×10^{-14} at night and 2×10^{-14} during the day.

References

1. J. Accetta. 1993. Infrared search and track systems. In *Passive Electro-Optical Systems,* vol. 5, ed. S. Campana, of *The Infrared and Electro-Optical Systems Handbook,* executive ed. J. Accetta and D. Shumaker. Ann Arbor, MI: ERIM, and Bellingham WA: SPIE, 287.
2. R. Beland. 1993. Propagation through atmospheric optical turbulence. In *Atmospheric Propagation of Radiation,* vol. 2, ed. F. Smith, of *The Infrared and Electro-Optical Systems Handbook,* executive ed. J. Accetta and D. Shumaker, Ann Arbor, MI: ERIM, and Bellingham WA: SPIE, 221.

FRIED PARAMETER

Subject: Atmospherics, lasers, astronomy, systems

The Rule

Fried has developed a useful characterization of the atmosphere. He has computed a characteristic scale of an atmospheric path, commonly referred to as r_0 and known as Fried's parameter, that has the property that a diffraction limited aperture of diameter r_0 will have the same angular resolution as that imposed by atmospheric turbulence, r_0 is computed as follows:

$$r_0 = \left(0.423 k^2 \sec^2 \beta \int_O^L C_n^2(z)\, dz \right)^{-3/5}$$

where k = propagation constant of the light being collected; $k = 2\pi/\lambda$
     β = zenith angle of the descending light waves
     L = path length through which the light is collected
     C_n^2 = atmospheric refractive structure function, discussed in the rule called
         " C_n^2 Estimates"
     λ = wavelength
     sec = secant function (the reciprocal of the cosine function)
     z = a dummy variable that represents the path over which the light propagates

Clearly, for any path over which C_n^2 is constant we get

$$r_0 = (0.423\ k^2 \sec^2 \beta C_n^2 L)^{-3/5}$$

For a vertical path that includes the typical profile of C_n^2, the value of r_0 is about 15 cm.

Basis for the Rule

Fried derived this expression using the Kolmogorov spectrum for turbulence. Continued development of the theory, particularly for the evaluation of the performance of space and aircraft remote sensors and for laser applications such as optical communication, has led to new approximations and more accurate characterizations of the impact of the atmosphere on propagation of light.

Cautions and Useful Range of the Rule

Proper characterization of C_n^2 is necessary to get a good estimate of Fried's parameter. Note also that there is a wavelength dependence in the results, hidden in the parameter k, which is equal to $2\pi/\lambda$. Unfortunately, characterization of C_n^2 is an imprecise empirical exercise. Astronomical sites measure it in a limited way, but it varies with location, season, and weather conditions, so it is usually only approximated. Of course, attention must be paid to using the correct units. Since C_n^2 is always expressed as $m^{-2/3}$, meters are the preferred units.

Usefulness of the Rule

This rule provides a convenient characterization of the atmosphere that is widely used in atmospheric physics, including communications, astronomy, and so on.

Properly applied, it provides a characterization of the maximum telescope size for which the atmosphere does not provide an impediment that blurs the spot beyond the diffraction limit. That is, a small enough telescope will perform at its design limit even if the presence of the atmosphere is taken into account.

The Fried parameter is often used in adaptive optics to determine the required number of active cells and the number of laser-generated guide stars necessary for some level of performance.

Fried's parameter continues to find other uses. For example, the resolved angle of a telescope can be expressed as approximately λ/r_o, or about 3.3 μrad for an r_o of 15 cm and a wavelength of 0.5 microns. Note that this result is consistent with those in the rule "Atmospheric Seeing."

Notes and Explanation

Convection, turbulence, and varying index of refraction of the atmosphere distort and blur an image, limiting its resolution. The best "seeing" astronomers obtain on good nights, at premier high-altitude observatories such as Mauna Kea, is on the order of 0.5 to 1.5 μrad. This limit is regardless of aperture size. It is not a question of diffraction limit but one of being "atmospheric seeing limited" due to atmospheric effects. The seeing tends to improve with increasing altitude and wavelength.

The Fried parameter is the radius in which the incoming wavefront is approximately planar. In the visible, it ranges from about 3 to 30 cm. The Fried parameter is strongly spatially and temporally dependent on the very localized weather at a given location, and it varies with the airmass (or the telescope slant angle). It also can be affected by such localized effects as the telescope dome and air flow within the telescope. Moreover, the Fried parameter can vary across the aperture of a large telescope. The parameter (hence, seeing) increases with wavelength so that, on good nights, moderate-size telescopes (say less than 5 m in aperture), operating at 10 μm or longer, tend to be diffraction limited. Stated another way, the Fried parameter in those cases exceeds, or at least equals, the telescope aperture.

The amateur astronomer's 5- to 10-inch aperture telescope is about as big as a telescope can be before atmospheric effects come into play, if the local environment (city lights, etc.) is not a factor. The really large telescopes have the same angular resolution, because they too are affected by the atmosphere. Of course, the big telescopes are not in your back yard but, rather, are sited where atmospheric effects are as insignificant as possible. In addition, large telescopes collect light faster than smaller ones and thus allow dimmer objects to be seen in a reasonable length of time.

One method to correct for this atmospheric distortion is to employ a wavefront sensor to measure the spatial and temporal phase change on the incoming light, and to use a flexible mirror to remove the distortions that are detected, in essence removing the atmospheric effects in real time. The wavefront sensor can be a Shack-Hartmann sensor, which is a series of lenslets (or subapertures) that "sample" the incoming wavefront at the size of (or smaller than) the Fried parameter. The size of the wavefront sensor subaperture and the spacing of the actuators that are used to deform an adaptive mirror can be expected to be less than the Fried coherence cell size. The diameter of the telescope divided by the Fried parameter indicates the minimal number of subapertures needed. The optimum size of the wavefront spacing and correction actuators seems to be between 0.6 and 1.0 times the Fried cell size. More details on this topic are covered in other rules in this chapter.

Source

C. Aleksoff et al. 1993. Unconventional imaging systems. In *Emerging Systems and Technologies,* vol. 8, ed. S. Robinson, of *The Infrared and Electro-Optical Systems Handbook,* executive ed. J. Accetta and D. Shumaker. Ann Arbor, MI: ERIM, and Bellingham WA: SPIE, 132.

ATMOSPHERIC TRANSMISSION AS A FUNCTION OF VISIBILITY

Subject: Atmospherics, transmission

The Rule

Atmospheric transmission can be estimated via the range, visibility, and wavelength by

$$\tau = \exp\left[\frac{-3.91}{V}\left(\frac{\lambda}{0.55}\right)^{-q} R\right]$$

where V = visibility in the visual band in km
   λ = wavelength in microns
   q = size distribution for scattering particles; typical values are 1.6 for high visibility, 1.3 for average visibility, and $0.585\,V^{1/3}$ for low visibility
   R = range in km

Basis for the Rule

Although there is a growing body of theoretical work that supports these types of theories, this result comes from empirical analysis and observations.

Cautions and Useful Range of the Rule

As can be seen, the choice of q depends on the visibility for each particular situation. Of course, this requires that the visibility be known. Furthermore, the rule assumes that the visibility is constant over the viewing path. This never happens, of course. Nonetheless, this is a useful and easy-to-compute rule that can be used with considerable confidence. Visibility is easily obtained from the FAA or National Weather Service and is readily available at most electro-optical test sites.

Usefulness of the Rule

This rule provides a quick and easy approach for estimating the transmission of the atmosphere as a function of wavelength if some simple characteristics are known. It is useful for field work related to lasers, observability of distant objects, and the applicability of telescopes, if visibility is known. Often, visibility is made available by the National Weather Service. Of course, one can attempt to estimate the visibility if transmission can be measured.

Notes and Explanation

All sorts of human enterprises involve looking through the atmosphere. Simple rules for establishing how far one can see have always been of interest. This little rule mixes a little physics, in the form of the size distribution effects, into the empirical transmission. The longer the wavelength, the less the scatter, so wavelength is in the numerator. As in another rule, the absorption can be estimated from the visibility by $4/V$.

  In general, aerosols have particle radii of 0.01 to 1 µm with a concentration of 10 to 1000 particles/cc, fog has radii of 1 to 10 µm with a concentration of 10 to 100/cc, clouds have particle radii of 1 to 10 µm with a concentration of 10 to 300 /cc, and rain has particle radii of 100 to 10,000 µm with a concentration of 0.01 to 10^{-5}/cc.[1]

Reference

1. M. Thomas, D. Duncan. 1993. Atmospheric transmission. In *Atmospheric Propagation of Radiation*, vol. 2, ed. F. Smith, of *The Infrared and Electro-Optical Systems Handbook*, executive ed. J. Accetta and D. Shumaker. Ann Arbor, MI: ERIM, and Bellingham, WA: SPIE, 12.

Additional Sources

P. Kruse, L. McGlauchlin, and R. McQuistan. 1962. *Elements of Infrared Technology*. New York: John Wiley & Sons, 189–192.

D. Wilmot et al. 1993. Warning systems. In *Countermeasure Systems*, vol. 7, ed. D. Pollock, of *The Infrared and Electro-Optical Systems Handbook*, executive ed. J. Accetta and D. Shumaker, Ann Arbor, MI: ERIM, and Bellingham, WA: SPIE, 31.

BANDWIDTH REQUIREMENT FOR ADAPTIVE OPTICS

Subject: Optics, adaptive optics, phenomenology, and environment

The Rule

To correct phase fluctuations induced by the atmosphere, an adaptive optics servo system should have a bandwidth of

$$\frac{0.4 v_w}{\sqrt{\lambda L}}$$

where v_w = wind velocity in meters/second
  λ = wavelength in meters
  L = path length inducing the phase fluctuations in meters

Basis for the Rule

Empirical observations

Cautions and Useful Range of the Rule

- Wind velocities of normal speed (0 to 20 km/hr)
- Normal wavelengths (visible to mid-IR)

Usefulness of the Rule

- Estimating the control bandwidth
- Estimating the maximum wind in which an adaptive system can operate

Notes and Explanation

This handy relationship indicates that the shorter the wavelength and the higher the wind velocity, the faster the servo system needs to be controlled. The bandwidth is lowered as the path length increases because of the effect of averaging over the path.

Source

R. Tyson. 1991. *Principles of Adaptive Optics.* Orlando, FL: Academic Press, 36.

NUMBER OF ACTUATORS IN AN ADAPTIVE OPTIC

Subject: Optics, adaptive optics, atmospherics

The Rule

To correct for atmospheric turbulence effects, an adaptive optic system needs a minimum number of actuators. The number of actuators required, if evenly spaced over the adaptive optic, is approximately

$$\left(\frac{\text{Telescope aperture}}{r_0} \right)^2$$

where r_0 = the form of Fried's parameter used for spherical waves:

$$3.024 \, (k^2 C_n^2 L)^{-3/5}$$

For plane waves, such as are received from star light,

$$r_0 = 1.68 \, (k^2 C_n^2 L)^{-3/5}$$

where L = distance the light propagates through the disturbing atmosphere

$k = 2\pi/\lambda$

C_n^2 = the atmospheric structure coefficient, which $\approx 10^{-14} \, m^{-2/3}$

λ = wavelength

Basis for the Rule

These results are derived directly from the turbulence theory of the atmosphere as has been described throughout this chapter. Considerable effort has gone into confirming the accuracy of the theory, as described in the introduction to this chapter and several of the rules.

Cautions and Useful Range of the Rule

The results shown here are for the simplifying situation in which the atmosphere is assumed to be constant over the path through which the light propagates. It also assumes that r_0 is smaller than the aperture that is equipped with the adaptive optics technology. We also note that the typical astronomical case involves correcting for the turbulence in a nearly vertical path through the atmosphere. The descriptions above for Fried's parameter apply only for constant atmospheric conditions. When the properties of C_n^2 (h) are known, the equations in the rule "Fried's Parameter" can be used to compute r_0. The complexity of computing r_0 for the nearly vertical case can be avoided by assuming that r_0 is about 15 cm.

Usefulness of the Rule

- Estimating the number of actuators required
- Estimating the minimum Fried parameter that a given adaptive optic can accommodate

- Also shows the wavelength dependence of the performance of an adaptive optics system. The wavelength dependence of the number of actuators goes as $\lambda^{-12/5}$, so longer wavelengths require fewer adaptive elements, as expected.

Notes and Explanation

As expected, the number of actuators depends on the properties of the atmosphere, the length of the path the light travels, the wavelength of the light and the application of the adaptive optics. The latter point derives from whether the light is a plane wave, such as pertains to star light, or spherical waves, such as characterize light beams in the atmosphere.

To properly compensate for atmospheric turbulence, the number of actuators depends on the above form of the Fried parameter and the size of the optic. The optical surface must be divided into more moveable pieces than the maximum amount of turbulent cells that can fit on the same area. If fewer actuators are used, then the atmosphere will cause a wavefront error that cannot be compensated for by the optic. Tyson[1] shows that a more accurate representation of the number of actuators is approximately

$$\left(\frac{0.05 \ k^2 L C_n^2 D^{5/3}}{\ln \ (1/S)} \right)^{6/5}$$

where S = the Strehl ratio desired (The Strehl ratio is a commonly used performance measure for telescope optics that essentially defines how closely an optical system comes to performing in a diffraction-limited way.)

Example

We see that, using the simple form of the rule, a 1 m aperture operating at a location that typically experiences a Fried parameter of 5 cm will need 400 actuators. This is typical of the nearly vertical viewing associated with astronomical applications. At the same time, if we use Tyson's version of the rule, we obtain that 400 actuators will produce a Strehl ratio of about 0.89.

Reference

1. R. Tyson. 1991. *Principles of Adaptive Optics*. Orlando, FL: Academic Press, 259.

Additional Source

H. Weichel. 1988. *Laser System Design*. SPIE Class Notes. Bellingham, WA: SPIE, 144.

INCREASED REQUIREMENT FOR RANGEFINDER SNR DUE TO ATMOSPHERIC EFFECTS

Subject: Atmospherics, lasers, systems

The Rule

In weak turbulence, the required signal-to-noise ratio (SNR) of a rangefinder must be increased to overcome the added scintillation noise. The increase is about

$$E_{snr} = \exp\left[\sqrt{2}\sigma_I \, \text{erf}^{-1} (2P_d - 1) + \frac{1}{2}\sigma_I^2 \right]$$

where E_{snr} = required increase to the SNR (required signal to noise in turbulence = $E_{snr} \times$ the required signal to noise in calm air)

P_d = probability of detection requirement

erf⁻¹ = inverse error function, which is defined in any number of books on advanced engineering, mathematics, or statistics, and is simplified in the "erf rule"

σ_I = variance of log intensity ratio (Additional information on the computation of I is found in the rule "Laser Beam Scintillation," on page 32.)

Basis for the Rule

This rule results from analysis of the ability of an active tracker to generate the necessary SNR, taking into account the fluctuating intensity at the target and the receiver due to the turbulence along the path.

Cautions and Useful Range of the Rule

This rule does not include the attenuation effects that also occur in the atmosphere due to particulate scattering and absorption. Those effects are rather easily included in the calculation. Attenuation properties of the atmosphere are discussed in the rule "Beer's Law," on page 13.

Usefulness of the Rule

This rule provides a simple explanation of the impact of turbulence on laser radar systems. System engineers will find it useful in assessing the impact of atmospheric effects and estimating system performance.

Notes and Explanation

Turbulence broadens a laser beam. In addition, it causes the center of the beam to meander around. This combination of broadening and beam wander causes the energy to be distributed over a larger angle and area than when the atmospheric effect is not present. This means that less energy is put on the target, thus reducing the signal reflected to the sensor. Therefore, the SNR will be less in turbulent conditions than in calm conditions. To achieve the same probability of detection and probability of false alarm, the signal-to-noise ratio must be increased.

Example

The following examples show how this rule can easily be used to determine the impact of the atmosphere on the performance of a laser radar system.

For wavelengths associated with doubled YAG laser light (0.532×10^{-6} m) operating at near ground level, C_n^2 is about 10^{-14} at night and 1.7×10^{-14} during the day. For light turbulence,

$$\sigma_I = 2\sigma_\chi$$

σ_χ where is computed from

$$\sqrt{0.31}\left(\frac{2\pi}{\lambda}\right)^{7/12} L^{11/12} C_n$$

Thus, for a night condition,

$$\sigma_I = 1.5 \times 10^{-3} L^{11/12} \quad \text{for } C_n = 1 \times 10^{-7}$$

For a nominal path length of 600 m, the signal intensity has a variation with a standard deviation of about 53 percent.

To achieve a probability of detection of 0.99, we complete the remainder of the terms in the equation.

$$\mathrm{erf}^{-1}(2 \times 0.99 - 1) \approx 1.65$$

Using the equation for this rule, we find that the enhancement in SNR must be 3.96. That is, the radiometrics of the system must considered to ensure that the combination of laser power and receiver sensitivity leads to an increase in SNR of nearly 4. During the day, C_n^2 is about 1.7×10^{-14}, and the enhancement requirement jumps to 6.4. This required increase is almost completely eliminated if the wavelength of the laser is 1.06 microns. This is because σ_χ depends on the inverse of the wavelength.

Source

R. Byren. 1993. Laser rangefinders. In *Active Electro-Optical Systems*, vol. 6, ed. C. Fox, of *The Infrared and Electro-Optical Systems Handbook*, executive ed. J. Accetta and D. Shumaker. Ann Arbor, MI: ERIM, and Bellingham, WA: SPIE, 103.

THE PARTIAL PRESSURE OF WATER VAPOR

Subject: Atmospherics

The Rule

A number of rules related to *upwelling* and *downwelling* radiation in the atmosphere depend on knowing the partial pressure of water vapor. Upwelling and downwelling describe the direction of flow of radiation. Sunlight is downwelling, while reflections from the surface are upwelling. The following rule matches the observed distribution by fitting data with a least squares curve. The resulting equation shows the partial pressure as a function of air temperature and relative humidity.

$$P = 1.333 \, R_h \{[(C_3 \, T_c + C_2) \, T_c + C_1] \, T_c + C_0\}$$

where P = partial pressure of water given in millibars
R_h = relative humidity
C_0 = 4.5678
C_1 = 0.35545
C_2 = 0.00705
C_3 = 3.7911 × 10^{-4}
T_c = air temperature in °C

Basis for the Rule

This rule is the result of empirical observations.

Cautions and Useful Range of the Rule

This rule applies for temperatures in the range from 0 to 30° C.

Usefulness of the Rule

The partial pressure is useful in a number of applications. For example, the same reference shows that the downwelling radiance onto a horizontal plane is proportional to the square root of the partial pressure of water vapor.

Notes and Explanation

This rule gives immediate results for estimating the partial pressure of water vapor. It clearly shows that, as the temperature rises, the partial pressure increases rapidly. In fact, review of the equation shows that the increase goes as T^3 for one term and T^2 for another.

Source

D. Kryskowski and G. Suits. 1993. Natural sources. In *Sources of Radiation*, vol. 1, ed. G. Zissis, of *The Infrared and Electro-Optical Systems Handbook*, executive ed. J. Accetta and D. Shumaker. Ann Arbor, MI: ERIM, and Bellingham, WA: SPIE, 147.

PULSE STRETCHING IN SCATTERING ENVIRONMENTS

Subject: Atmospherics, lasers

The Rule

The increase in transmitted laser pulse duration due to scattering in the atmosphere can be estimated as

$$\Delta t = \frac{L}{c} \left\{ \frac{0.3}{a\tau\theta_{rms}^2} \left[\left(1 + 2.25\, a\tau\theta_{rms}^2 \right)^{1.5} - 1 \right] - 1 \right\}$$

where L = propagation distance

c = speed of light

a = single scatter albedo ≈ 1

τ = product of the scatter cross section per unit volume and the propagation distance (τ ranges from 12 to 268 in the experiments illustrated in the paper, and it is unitless.)

θ_{rms} = rms scatter angle $\approx 30°$ for water (θ_{rms} is expressed in radians in the equation.)

Basis for the Rule

This result comes from analysis of the multiple scattering that occurs in the atmosphere, using some simplifying assumptions. In addition, the formulation uses some results from the classical theory of electron scattering of electromagnetic waves.

Cautions and Useful Range of the Rule

This analysis assumes that the rules of geometric optics apply. That is, it assumes that diffractive and other complex effects that lead to phase distortion in the beam are not present. Furthermore, it is likely that the rule breaks down for optical depths (τ) in excess of about 300, so it will not work well for dense fogs and clouds.

Usefulness of the Rule

Those involved in laser communications in the atmosphere will find this rule helpful because pulse stretching limits the data rate of the channel. This becomes evident when we consider situations in which the last photons to arrive from a first pulse are still arriving, having gone through many multiple scattering paths, as the first light arrives from a second pulse. Clearly, this would confound the receiver and prevent it from properly interpreting the data.

Notes and Explanation

Stotts[1] shows that this formulation compares well with both the experiment and more complex simulations using Monte Carlo methods. In view of the simplicity of the rule, this is a most attractive place to start. Systems that require more accuracy can use the Monte Carlo methods referenced in Stotts' paper. The results in the reference show that pulse stretching on the order of microseconds results from situations in which fog or clouds are present. This will have a profound impact on the function of a pulsed communication system.

Reference

1. L. Stotts. 1978. Closed form expression for optical pulse broadening in multiple-scattering media. *Applied Optics,* Volume 17, No. 4.

LASER BEAM SCINTILLATION

Subject: Atmospherics, lasers, propagation of light

The Rule

Variance in the irradiance from a beam due to turbulence can be estimated as follows for the special case of a horizontal beam path. Variance in irradiance is defined as

$$\sigma^2 = 4\sigma_l^2$$

where $\sigma_l^2 = 0.31\ C_n^2\ k^{7/6} L^{11/6}$

$\qquad k = 2\pi/\lambda$

$\qquad C_n^2$ = index of refraction structure constant discussed in the rule called "C_n^2 Estimates"

$\qquad L$ = the propagation path length in the turbulent medium

$\qquad \lambda$ = wavelength

Combining the equations above, we get

$$\sigma^2 = 1.24\, C_n^2\ k^{7/6} L^{11/6}$$

The standard deviation variation in irradiance is the square root of the term on the right.

Basis for the Rule

This type of simplifying analysis of beam propagation in the atmosphere has been spearheaded by both the military, who are interested in laser beam propagation, and the astronomical community, who are very much concerned with the disturbance that the atmosphere imposes on light collected by terrestrial telescopes. Of course, the latter group has little use for analysis of horizontal propagation of light, but the underlying theory that results in the equations above can be applied to the vertical views that astronomers prefer if the refractive properties of the atmosphere are known.

Cautions and Useful Range of the Rule

A basic assumption used in the development of the results for the horizontal beam case is that the value of C_n^2 is constant over the path. This is generally not the case since C_n^2 is the manifestation of temperature variations in the atmosphere. However, this simplifying assumption is often used for horizontal beams and has been qualitatively confirmed in many experiments. In the case that C_n^2 varies along the path, a more complicated formalism must be used in which various $C_n^2\ (h)$ models are employed, as described elsewhere in this chapter.

The above equations apply to plane waves. Spherical waves can be characterized by the same analysis, except that the first equation uses 0.124 as the multiplier rather than 0.31. In addition, there is a limit to the range of atmospheric conditions over which the rule applies. The best estimate is that the expressions can be used if σ_l^2 is not bigger than about 0.3.

Usefulness of the Rule

Use this rule to make rapid assessments of the performance of laser beam and other light transmission systems in the presence of atmospheric effects. The reference also provides the additional details necessary to deal with beam paths that are not horizontal. To do so requires that C_n^2 be known or estimated as a function of altitude.

Notes and Explanation

Use other rules in this chapter to estimate C_n^2 as a function of altitude.

Example

For example, if we use a typical C_n^2 value of 10^{-14}, a path length of 375 m and a wavelength of 0.5 microns, we get an irradiance variance of about 0.12, which is equivalent to a standard deviation of about 35 percent variation in the intensity at the receiver.

Source

J. Accetta. 1993. Infrared search and track systems. In *Passive Electro-Optical Systems,* vol. 5, ed. S. Campana, of *The Infrared and Electro-Optical Systems Handbook,* executive ed. J. Accetta and D. Shumaker. Ann Arbor, MI: ERIM, and Bellingham WA: SPIE, 288.

LASER BEAM SPREAD

Subject: Atmospherics, lasers, systems

The Rule

The angular spread of a beam along an atmospheric path is[1]

$$\theta^2 = \frac{1}{k^2 a^2} + \frac{1}{k^2 \rho^2}$$

where $k = 2\pi/\lambda$
a = beam radius

The parameter ρ is the transverse coherence distance and is related to the effect that the turbulence of the atmosphere has on propagation of light.

$$\rho = \left[1.46 \, k^2 \int_0^L C_n^2(z) \left(\frac{z}{L}\right)^{5/3} dz \right]^{-3/5}$$

where L = path length
C_n^2 = refractive index structure constant, as further defined in the C_n^2 rule
z = a dummy path variable that covers the path over which the light travels

Although this equation looks daunting, remember that k is a constant, and C_n^2 may be as well. Even if C_n^2 changes along the path, you can compute the integral with a calculator if you keep your wits about you.

Basis for the Rule

This rule comes from a combination of the analytic description of laser beam propagation in a vacuum along with a simplified assumption about the way that other beam spreading effects, such as caused by the atmosphere, add to the theoretical beam spread.

Cautions and Useful Range of the Rule

These estimates of beam size are as good as the quality of the estimates of C_n^2, since the description of beam spreading, derived from the theory of laser resonators, has been proven to be exact enough to predict all of the key features of laser radiation.

Also note that this rule does not include the effects of particular light scattering by the atmosphere.

Finally, it should be noted that various authors use two different descriptions for ρ. One is for plane waves and the other is for spherical waves. The expression for ρ as used above is the spherical wave case, which is the typical form used for laser beams propagating in the atmosphere. The plane wave case is typically only used for star light propagating in the atmosphere.

Usefulness of the Rule

This rule provides a quick and easy estimate of the size of a beam as a function of distance from the laser under the influence of the atmosphere. It is likely to be useful to those designing or modeling laser communications or LIDAR systems.

Notes and Explanation

A special case pertains for a horizontal beam path.

$$\rho = \left[1.46 \left(\frac{3}{8} \right) k^2 \, C_n^2 \, L \right]^{-3/5}$$

for a horizontal beam.[2] This derives directly from the previous definition of ρ as presented above.

C_n^2, defined in the rule called "C_n^2 Estimates," is about 1×10^{-14} near the ground and 1×10^{-18} at 20 km altitude.

Tyson[2] also reports that, for short propagation distances, a horizontal laser beam spreads to a waist size w_b, expressed as

$$(w_o^2 + 2.86 \, C_n^2 \, 2k^{1/3} L^{8/3} w_o^{1/3})^{1/2}$$

and over longer distances, $L \gg \pi w_o^2 / \lambda$

$$w_b^2 = \frac{4L^2}{k^2 w_o^2} + 3.58 \, C_n^2 L^3 w_o^{-1/3}$$

In these equations, w_o refers to the beam size as it exits the laser.

Laser beams are used for so many applications that the study of their beam spread in turbulence is of considerable use. Clearly, the simplest results are obtained when one can accurately assume that C_n^2 is constant. This is rarely the case, but the assumption is adequate for many applications. Considerable attention has been paid to this problem by a number of researchers. Recently, the emphasis has been in three areas: military applications, communications based on pulsed laser beams, and astronomical telescope systems in which a laser beam is used to create a synthetic star that is used to provide information on corrections that must occur to remove the turbulence of the atmosphere.

References

1. R. Tyson. 1991. *Principles of Adaptive Optics.* San Diego: Academic Press, 32.
2. R. Tyson and P. Ulrich. 1993. Adaptive optics. In *Emerging Systems and Technologies,* Volume 8, ed. S. Robinson, of *The Infrared and Electro-Optical Systems Handbook,* executive ed. J. Accetta and D. Shumaker. Ann Arbor, MI: ERIM, and Bellingham, WA: SPIE, 180.

3

Acquisition, Tracking, and Pointing

Acquisition, tracking, and pointing (ATP) are traditional functions in a number of scientific and military enterprises and systems. Recent developments have vastly increased this capability. One can imagine early hunters going through the same activities as today's heat seeking missiles when the concepts of ATP are applied. For instance, the hunter looks for a bird to kill for dinner and, like the fighter pilot, eventually finds prey. Acquisition takes place as all other distractions are eliminated and attention is turned to a particular target. Next, the brain of the hunter and the computer in the aircraft, begin to observe and follow the flight of the victim. Assuming that random motions aren't employed, which is a protection tactic for both types of prey once they know they are being considered as a target, the tracking function begins to include anticipation of the trajectory of the bird/target. This anticipation is critical for allowing for "lead ahead" aiming since the weapon can't be pointed at the current position, but rather must be pointed at the collision point, if there is to be a collision. Finally, the hunter must aim (or point) his weapon properly, taking into account the lead ahead effect, as well as the size of the target. Today's technology solves the size problem by constantly tracking and correcting the trajectory of the weapon so that the closer it gets, the more accurate the hit.

The same type of function occurs at much lower rates in a variety of scientific applications of EO. For example, astronomical telescopes must acquire and track the stars, comets and asteroids they intend to study. The rate at which the star appears to cross the sky is much lower than that of a military or culinary target but the pointing must be much more accurate if quality images are to be obtained. There are few EO systems in which at least one part of the triad of ATP isn't hard to do!

Of course, the great leaps forward in ATP over the last century have derived from a combination of optical technologies and electronics. Some advances derive, in theory and practice, from the previous development of RADAR during WWII. Indeed, some of the rules in this chapter relate to the computation of the probability of detection of targets. These computations have direct analogs in both the EO and

RADAR worlds. A common features of both RADAR and EO tracking systems is the concern about the signal-to-noise ratio that results from different types of targets. This is because the quality of the ATP functions depend in a quantitative way on the size of the SNR. Targets that are not well distinguished from the background and noise sources inside the instrument will be poorly tracked and eventual success of the system will be compromised, even if the pointing system is very capable.

Through it all, the development of ATP systems rests on several basic principles; the shape and size of the target, the range of the target, the contrast between the target surface characteristics and those of the surrounding scene, the intensity of the target signature, the ability of the system operator to point the weapon and the speed at which it can accommodate the trajectory (crossing rate) of the target. Some modern implementations of these systems rely on advanced mathematical algorithms for estimating the trajectory, based on the historical motions of the target. They also rely heavily on the control systems that will point both the tracking system and the weapon. It is clear from recent successes in both military and scientific systems that ATP is a mature technology that is becoming limited only by the environment through which the system must view the target.

This chapter covers a number of rules related to EO sensing and detection, but it also addresses some empirical observations related to how humans perform detection and tracking functions. Our ancestors didn't know it, but they were using the same type of rules that appear in "Target Resolution vs. Line Pair Rules" (page 48) in which quantitative expression is given to the resolution needed for performing the acquisition, tracking, and pointing functions.

SIMPLE TARGET DETECTION RULES

Subject: ATP, systems

The Rule

There are four typical situations encountered by the electro-optical (E-O) designer. The ranges at which effective tracking can be performed are qualitatively described by the table below. Back-of the envelope estimates usually give reliable answers to two of them. Performance for the other two cases depends significantly on the way the design is carried out and on subtleties of the processing, mission use, and innovated exploitation of target phenomenology. Expect extra effort to obtain a version that works in these cases.

Geometry	Thrusting target	Nonthrusting target
Target viewed above the horizon	Tracking range $\approx \infty$	It depends
Target viewed below the horizon	It depends	Tracking range ≈ 0

Basis for the Rule

This is based on empirical observations and generalization of mediocre engineering.

Cautions and Useful Range of the Rule

This is a generalization, so a detailed analysis should always be performed.

Some very small thrusting targets (such as shoulder-launched missiles and post boost vehicles) will still present problems at long ranges, even if they are viewed with a sky background, and particularly if the sun is anywhere near the field of view.

Usefulness of the Rule

This rule is really most effective in reminding system designers not to try to do something that is inherently quite difficult. At a minimum, it suggests situations in which there will be a significant challenge.

Notes and Explanation

One can crudely divide targets into two categories: thrusting and nonthrusting. Thrusting targets are easy to see, and nonthrusting targets (unfortunately most of the world is nonthrusting) are sometimes difficult to detect. Likewise, backgrounds can be grossly divided into two categories: above the horizon and below the horizon. Generally, targets above the horizon present a low background without much clutter, and targets below the horizon present a warm and cluttered background.

SNR Requirements

Subject: ATP, systems, backgrounds

The Rule

A signal-to-noise ratio (SNR) of 6 is adequate to perform all sensing and tracking functions. Any more SNR is not needed. Any less will not work. Targets can be sensed with certainty to a range defined by the SNR and allowable false alarm rate. Beyond that range, they are not readily detectable.

Basis for the Rule

This rule derives directly from standard results in a variety of texts that deal with target detection in noise. See the rule "P_d Estimation" in this chapter for additional details on how to compute the probability of detection for a variety of conditions.

Cautions and Useful Range of the Rule

Clearly, there are some cases in which the probability of false alarm (P_{fa}) can be raised, which allows a small drop in the SNR for a required probability of detection (P_d). For example, if one is willing to tolerate a P_{fa} of 1 in 100, then the requirement on SNR drops from 6 to about 4 for 90 percent P_d.

This rule assumes "white noise", in which noise is present in all frequencies with the same probability. The situation is different with noise concentrated at a frequency or with certain characteristics. In general, if you have *a priori* knowledge of the characteristics, you can design a filter to improve performance. However, if you do not know the exact characteristics (which is usually the case with clutter), then your performance will be worse than expected by looking at tables created for white noise. The more complex case in which the noise has some "color" cannot be dealt with so easily because all sorts of possible characterizations can occur.

Usefulness of the Rule

This rule can guide the designer or system analyst to avoid overdoing the design or ending up with a design that will not work at the desired range. Additional SNR may be required for subpixel tracking.

Notes and Explanation

Another rule in this chapter provides a short-cut method of computing P_d. It is called "P_d Estimation."

Figure 3.1 illustrates the high probability of detection levels for an SNR of 6. Also, note that the probability of detection increases rapidly as SNR increases (at a false alarm of 10^{-4}. Doubling the SNR from 3 to 6 results in a probability of detection to increase from about 0.1 to well above 0.95).

Example

For example, the *Burle Electro-Optics Handbook*[1] shows that a probability of detection in excess of 90 percent can only be achieved with a probability of false alarm of around one in a million if the SNR is about 6. In most systems, a P_{fa} of at most one in a million is about right. It must also be noted that the data shown in the reference are pixel rates. Therefore, in a large focal plane, there may be around one million pixels and, as a result, the P_{fa} requirement of approximately one in a million limits the system to one false alarm per frame.

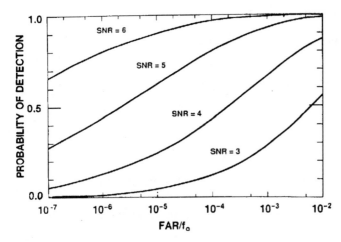

Figure 3.1 Probability of detection levels

Reference

1. Burle Industries. 1974. *Burle Electro-Optics Handbook*. Lancaster, PA: Burle Industries, 112.

Additional Source

D. Wilmot et al. Warning systems. In *Countermeasure Systems*, vol. 7, ed. D. Pollock, of *The Infrared and Electro-Optical Systems Handbook*, executive ed. J. Accetta and D. Shumaker, Ann Arbor, MI: ERIM, and Bellingham, WA: SPIE, 61.

PROBABILITY OF DETECTION ESTIMATION

Subject: ATP, systems, performance estimation

The Rules

There are some simple expressions that provide very good approximations to the probability of detection, such as follows:[1]

$$P_d \approx \frac{1}{2}\left[1 + \mathrm{erf}\left(\frac{l_s - l_t}{\sqrt{2}\,l_n}\right)\right]$$

where $I_{s,t,n}$ = the detector current associated with s (signal), t (threshold) and n (noise)

In addition, Kamerman[2] has a different approximation, as follows:

$$P_d \approx \frac{1}{2}\left\{\left[1 + \mathrm{erf}\left(\sqrt{\frac{1}{2} + SNr}\right) - \sqrt{l_n\left(\frac{1}{P_{fa}}\right)}\right]\right\}$$

for SNR > 2 and P_{fa} between 10^{-12} and 10^{-3}

The reader may find it useful to use an approximation of the erf offered by Press et al.[3] This source shows that the complementary error function [erfc(x)] may be expressed as a polynomial approximation as follows:

erfc(x) = Texp($-Z^2$ – 1.26551223 + T(1.00002368 + T(0.37409196 + T(0.09678418
+ T(–0.18628806 + T(0.27886807 + T(–1.13520398 + T(1.48851587
+ T(–0.822152223 + T(0.17087277)))))))))

where Z = the absolute value of x
T = 1 / (1 + 0.5 Z)
erf(x) = 1 – erfc(x)

Press states that the above polynomial expansion expression is "based on Chebyshev fitting to an inspired guess as to the functional form."

Basis for the Rule

The detection of targets is described in mathematical terms in a number of contexts. However, there is commonality among the formulations used for radar, optical tracking, and other applications. The above equations are based on theoretical analysis if detection of point sources in white noise and backed up by empirical observations. Calculation of the probability of detection, P_d, of a signal in Gaussian noise is, unfortunately, quite complex. The exact expression of this important parameter is:[2]

$$P_d = \frac{1}{\pi}\int_{V_T}^{\infty} x\,\exp\left(-\frac{x^2 + A^2}{2}\right)\left[\int_0^p \exp\left(xA\,\cos y\right)dy\right]dx$$

where x and y = integration variables

$$A = \sqrt{2 \text{ SNR}}$$
$$V_T = \sqrt{-2 \ln (P_{fa})}$$
P_{fa} = probability of false alarm
SNR = signal-to-noise ratio

Cautions and Useful Range of the Rule

As in any other approximation, the limits of application of the simplified version must be considered before using the rule. However, the rules shown above are quite broad in their application range. For example, the version taken from Kamerman applies for SNRs above 2 and P_{fa} between 10^{-12} and 10^{-3}. For almost any application, the SNR will be 5 or greater, and the P_{fa} of 10^{-3} is about the greatest that is considered.

This rule assumes white noise, which means that the noise present in the system has equal amplitude at all temporal frequencies. This approximation is required to develop any general results because the spectral characteristics of noise in real systems are more complex and can, of course, take on an infinite number of characteristics.

It also assume point source detection.

The polynomial expansion is useful in an SNR range of 1 to 10, and Press states that it is accurate to within 1.2×10^{-7}.

Usefulness of the Rule

This rule makes the computation of P_d easier for the specified circumstances (e.g., white noise). In fact, using a simplified expression for the erf function, even simple spreadsheet models of tracking performance can be implemented.

Notes and Explanation

The suggested approximations are quite adequate for a number of applications in spite of their simplicity. The same equations can be applied to radar and other systems as well if the parameters are properly interpreted.

The abbreviation "erf" refers to the error function, which is defined as follows:

$$\text{erf}(x) = \frac{2}{\sqrt{\pi}} \int_0^x e^{-u^2} du$$

The error function is widely available in reference tables and is included in most modern mathematical software. For example, it can be found in virtually any book on statistics or mathematical physics.

References

1. Burle Industries. 1974. *Burle Electro-Optics Handbook*. Lancaster, PA: Burle Industries, 111.
2. G. Kamerman. 1993. Laser radar. In *Active Electro-Optical Systems*, vol. 6, ed. C. Fox, of *The Infrared and Electro-Optical Systems Handbook*, executive ed. J. Accetta and D. Shumaker, Ann Arbor, MI: ERIM, and Bellingham, WA: SPIE, 45.
3. W. Press et al. 1986. *Numerical Recipes*. Cambridge, MA: Cambridge University Press, 164.

SENSOR COVERAGE

Subject: ATP, systems

The Rule

It is often desirable to ensure mission coverage using many smaller sensors rather than a few big, costly ones.

Basis for the Rule

The rule is based on empirical observations and the state of the art. This rule emphasizes the fact that all viewing media (even space) are dirty. Since super-sensitivity is required to perform coverage with a small number of sensors (as opposed to many less-sensitive ones located closer to their targets), an array employing fewer sensors will be more significantly affected by dust, paint flecks, and other close-range objects, to the detriment of long-range target detection.

Cautions and Useful Range of the Rule

This rule is a gross generalization, but it is clear that, in the case of orbital systems, a large number of small satellites in appropriate orbits can provide at least as much coverage as a few, much more expensive ones. As in all system studies, individual scenarios and requirements should be fully analyzed and traded.

Usefulness of the Rule

It underscores the danger of assuming more sensitivity is always better.

It underscores a basic property to consider in design trade-offs.

Notes and Explanation

Several small sensors can provide the same coverage as one big and expensive sensor. Often, the best plan is to employ many small sensors because of such considerations as producibility, down-time, learning curves, and expense. Furthermore, the larger number of sensors will usually result in systems in which the targets are closer to the sensor. This allows all aspects of the sensor and spacecraft to be smaller.

OVERLAP REQUIREMENTS

Subject: ATP, scanning, step-staring, systems

The Rule

In a step-stare (or step-scanned) pattern, the overlap from one step to another should be 50 percent, and never less than a minimum of about 5 or 10 percent.

Basis for the Rule

This rule is based on empirical observations.

Cautions and Useful Range of the Rule

The amount of overscan really depends on the accuracy of the scan to scan correlation.

Usefulness of the Rule

It is useful in estimating the amount of overscan a system should have.

Notes and Explanation

When a system employs a step-stare or scanning in more than one dimension, it is advisable to overlap the scan or steps. The amount of this overlap is determined by the step-to-step correlation and processing. In general, it is advisable to overlap each new step by 50 percent of the area of the previous step. This requirement provides for confident registration and coverage. However, in a very accurate system with inertial data, this can be reduced to a few percentage points. In such a system, post-processing will register the frames properly.

Another approach is the inclusion of line-of-sight control using a fast steering mirror-pointing system that relies on inertial data for mirror control or uses a reference point in or near the target to stabilize the pointing system.

Frequently, the image processor will require information from some pixels on the previous scan to properly execute its algorithms. Imagine a 7×7 spatial filter with its values being set by the surrounding 9×9 box. To properly execute such an algorithm for the edge pixel of a scan or step-stare, the 9 pixels of the previous scan must be known, either from overlap or memory.

DETECTION CRITERIA

Subject: ATP, detection, systems

The Rule Targets can be easily detected if half (or more) of the photons in the pixel are from that target and half are from other noise sources.

Basis for the Rule

The signal-to-noise ratio of a detector is

$$\frac{\bar{e}_s}{\sqrt{\bar{e}_s + \bar{e}_n}}$$

In this equation, $\bar{e}_s$ represents the number of electrons generated in the detector by signal photons. It is equal to the rate of arrival at the focal plane of signal photons times the quantum efficiency of the detector and the integration time. $\bar{e}_n$ is the number generated by all of the noise sources, and it is made up of contributions from photon noise in the background along with the contribution from internal detector noise. If $\bar{e}_s = \bar{e}_n$, the signal-to-noise ratio is

$$\frac{\bar{e}_s}{\sqrt{2\bar{e}_s}} \quad \text{or} \quad 0.71\sqrt{\bar{e}_s}$$

Because the generation of electrons from the signal is proportional to the arrival of photons from the target, and this number is generally quite large, the SNR can be quite high. This is rarely the result expected by the uninitiated.

Cautions and Useful Range of the Rule

It does not include the effects of clutter.

Signal-to-noise ratios must still be greater than about 5; therefore, more than 70 photons per integration time should fall on the detector (assuming a 70 percent quantum efficiency). If no noise sources are present, then this equation is reduced to

$$\sqrt{\bar{e}_s}$$

and only about 25 signal electrons are needed for detection.

Usefulness of the Rule

It is useful to quickly estimate whether a target can be detected when immersed in a given background or noise.

It illustrates (or reiterates) that you do not need more photogenerated electrons from your target than you have from your noise sources to obtain a healthy signal-to-noise ratio.

Notes and Explanation

Photons from noise sources in the scene generate electrons in direct proportion to the quantum efficiency at their wavelength. Simple electronic filters will reduce the noise from these sources to the square root of the number of these electrons, while the photogenerated electrons from the target (and clutter) pass through these simple filters unattenuated.

For laser detection, avalanche photodiode, UV, and visible systems, the noise is frequently dominated by shot noise from the target and background. Therefore, this holds true if the total of the target and background photogenerated electrons is on the order of 50 or more.

This also holds true for IR systems, as both their targets and backgrounds generate many more electrons.

TARGET RESOLUTION VS. LINE PAIR RULES

Subject: ATP, systems, phenomenology

The Rules

If the following number of line pairs, on a monochromic display, are found across the target dimension, human observers usually will be able to complete the task successfully.[1]

- Detection (an object is present): 1.0 ± 0.25 line pairs
- Orientation (the potential direction of motion of the object is known): 1.4 ± 0.35 line pairs
- Recognition (the type of object is known): 4.0 ± 0.8 line pairs
- Identification (type of vehicle is known): 6.4 ± 1.5 line pairs

Basis for the Rules

This is founded on empirical observations of actual likely users viewing tactical military targets. The functions of detection, recognition, and identification of a target by a human observer depend on the SNR and the accuracy with which the target is resolved. A number of empirical results have been obtained for a variety of targets.

Cautions and Useful Range of the Rules

The most important caution about this rule is that it assumes that the observer is relying on a single wavelength for the detection process. A polyspectral instrument need not be an imager to provide sufficient target detection exceeding the above specifications for selective targets.

These results are sensitive to test setup, human proficiency, and target class.

This rule does not include the deleterious effects of clutter or other atmospheric effects.

Resolution as described above will result in proper performance by 50 percent of the observers asked to do it under nominal conditions.

The rule assumes sufficient SNR, contrast, and spatial resolution on a display to easily resolve the said line pairs.

Usefulness of the Rules

This type of rule is useful in determining if a particular electro-optical design will aid a human observer viewing a display.

It is useful in setting system requirements and, in general, human-to-display system design.

It is useful in estimating system resolution requirements.

Notes and Explanation

More detail on the expected number of line pairs on a target is contained in Table 1, from Howe.[2]

Different researchers obtain differing results, indicating that the designer needs to be flexible to accommodate the characteristics of the system users, working conditions, and so forth. For instance, the identification of the *class* of ship being observed requires 400 to 1000 pixels, whereas the classification of *types* of ships requires 66 to 400. Conversely, if the information is fused with radar, hyperspectral, or other information, the number of line pairs is reduced.

TABLE 1 Expected Number of Line Pairs on a Target

Target	Detection	Orientation	Recognition	Identification
Truck	0.9	1.25	4.5	8.0
M48 tank	0.75	1.2	3.5	7.0
Stalin tank	0.75	1.2	3.3	6.0
Centurion tank	0.75	1.2	3.5	6.0
Halftrack	1.0	1.5	4.0	5.0
Jeep	1.2	1.5	4.5	5.5
Command car	1.2	1.5	4.3	5.5
Soldier	1.5	1.8	3.8	8.0

A line pair is a particular way to define spatial resolution. It is equal to a bar and a space, or one cycle across a target. Therefore, when 4.0 line pairs are quoted, it is equal to identifying a pattern with 4 bars and 4 spaces between them.

The above line pair specifications are across the target's "critical" dimension, which can be assumed to be the minimal dimension for a worst-case scenario.

For some history, in the late 1950s, John Johnson experimented with the effect of resolution on one's ability to perform target detection, orientation, recognition, and identification functions. This was followed by much research by Johnson, Ratchees, and others from the 1950s through the 1980s.

Finally, there has been much discussion and controversy about the actual number of line pairs needed to do the functions, depending on clutter, spectral region, quality of images, test control, and *a priori* knowledge. However, rarely have the numbers suggested herein appeared to be incorrect by more than a factor of 2.

References

1. Burle Industries. 1974. *Burle Electro-Optics Handbook*. Lancaster, PA: Burle Industries, 121.
2. J. Howe. 1993. Electro-optical imaging system performance prediction. In *Active Electro-Optical Systems*, vol. 4, ed. M. Dudzik, of *The Infrared and Electro-Optical Systems Handbook*, executive ed. J. Accetta and D. Shumaker, Ann Arbor, MI: ERIM, and Bellingham, WA: SPIE, 92, 99.

Additional Sources

B. Tsou. 1993. Systems design considerations for a visually coupled system. In *Emerging Systems and Technologies*, vol. 8, ed S. Robinson, of *The Infrared and Electro-Optical Systems Handbook*, executive ed. J. Accetta and D. Shumaker, Ann Arbor: ERIM and Bellingham Wa: SPIE., page 520–521.

Holst, G. 1995. *Electro-Optical Imaging System Performance*. Winter Park, FL: JCD Publishing, 412–440.

SUBPIXEL ACCURACY

Subject: ATP, systems

The Rule

One can determine the location of a blur spot on a focal plane (in focal plane coordinates) to an accuracy that equals the resolution divided by the approximate signal-to-noise ratio, or

$$A_{LOS} \approx \text{Constant} \frac{\text{Angular Limit}}{\text{SNR}}$$

where A_{LOS} = line-of-sight noise (or tracking accuracy)
Angular Limit = the larger of the diffraction limit, blur spot, or pixel footprint
      SNR = signal-to-noise ratio

The constant is usually approximated the following way: for initial analysis, the constant is 1; for more detailed analysis, a number of other values are used, such as are described later in this section.

Basis for the Rule

The rule is based on empirical performance of existing systems. In addition, a number of workers have calculated the theoretical limit of subpixel tracking using the method of centroids. In this analysis, it is assumed that the light from the target is projected onto a number of pixels in the focal plane, with the minimum being four for a staring system (scanning systems can use fewer pixels because the image can move over the detector). This means that the FPA performs like the quad cells used in star trackers and other non-imaging tracking systems. By measuring the light falling in each of the four cells, the centroid of the blur can be computed. The limit in performance of such systems is about 1/100 of the pixel IFOV, although higher performance has been demonstrated in some systems. Of course, the minimum pixel SNR must exceed about 5, or the target will not be reliably tracked.

Cautions and Useful Range of the Rule

This "superresolution" requires that the signal from each pixel in the focal plane be made available to a computer that can compute the centroid of the blur. In advanced systems, either electronic or optical line of sight stabilization is employed to ensure that each measurement is made with the blur on the same part of the focal plane. This eliminates the effect of noise that results from nonuniformity in pixel responsivity and noise. Results do not include transfer errors that build up between coordinates at the focal plane and the final desired frame of reference

Often, this subpixel accuracy does not work at an SNR of less than 5 or so, nor are full benefits realized at SNRs above 150 or so.

Usefulness of the Rule This rule is quite useful in determining the signal-to-noise ratio and pixel dimension required to ensure adequate tracking performance.

Notes and Explanation

A system can locate a target to an angular position of roughly the optical resolution of the system, divided by the SNR, for SNRs up to about 100. Beyond 100, saturation, crosstalk, blooming, and blurring effects usually limit the amount of subpixel information available.

In a scanning system, the blur circle produced by the target at the focal plane is scanned over the FPA element. The FPA element can be sampled faster than the time it takes the blur to move over the FPA. This provides a rise-and-fall profile to which the location can be calculated to an accuracy greater than the pixel footprint or blur extent. The higher the SNR, the faster the sample can be, and the more accurate the amplitude level will be, both increasing the accuracy of the rise-and-fall profile.

In a staring system, the target rarely falls centered on a given pixel. Usually, it is split between two or four pixels. Again, the signal on the adjacent pixels can be used to better define its location.

Several variants of the above rule are given below. From Held and Barry[1] we have

$$\frac{\pi 1.22\lambda}{8\ D\ \text{SNR}}$$

where λ = wavelength
$\quad\quad D$ = aperture diameter

This means that the constant referred to at the beginning of this article is equal to about 0.39. Also, Shao and Colavita[2] give

$$\frac{3\pi\lambda}{16\ D\ \text{SNR}}$$

Additionally, one of authors (Friedman) has computed the angular noise to be

$$\frac{0.31\lambda}{(D)\ \text{SNR}}$$

Stanton[3] has included position tracking error and defined it as

$$\theta_{\text{LSB}} = \frac{\text{pixel IFOV}}{\text{SNR}}$$

where θ_{LSB} = least significant bit of greatest resolution
$\quad$ pixel IFOV = instantaneous resultant field of view of a pixel

A scanning system has different equations for the cross-scan direction. Lloyd[4] points out that the accuracy for a cross-scan or when the only knowledge is that the target location falls within a detector angular subtense DAS is

$$\frac{\text{DAS}}{\sqrt{12}}$$

Finally, the measurement error in the angular separation of two objects is just like the rule stated above except that, if the SNRs are approximately the same, then

$$SNR = \frac{SNR_{1 \text{ or } 2}}{\sqrt{2}}$$

and the angular resolution is

$$\frac{\sqrt{2} \text{ pixel field of view}}{SNR}$$

References

1. K. J. Held and J. D. Barry. 1986. Precision optical pointing and tracking from spacecraft with vibrational noise. *Proc. SPIE,* vol. 616, Optical Technologies for Communication Satellite Applications.
2. M. Shao and M. Colavita. 1992. Long-baseline optical and infrared stellar interferometry. *Ann. Rev. Astron. Astrophs.* (30), 457–498.
3. C. Stanton et al. 1987. Optical tracking using charge coupled devices. *Proc. SPIE* 1489, 163–176.
4. J. Lloyd. 1993. Fundamentals of electro-optical imaging systems analysis. In *Active Electro-Optical Systems,* vol. 4, ed. M. Dudzik, of *The Infrared and Electro-Optical Systems Handbook,* executive ed. J. Accetta and D. Shumaker, Ann Arbor, MI: ERIM, and Bellingham, WA: SPIE, 42–44.

Additional Source

J. Miller. 1994. *Principles of Infrared Technology.* New York: Van Nostrand Reinhold, 60–61.

RANGE AS A FUNCTION OF THE NUMBER OF DETECTORS

Subject: ATP, atmospherics, systems

The Rule

The range performance of a sensor is dependent on the log of the number of detectors.

$$\text{Range} \propto \log_{10} N_d$$

where N_d = number of detectors

Basis for the Rule

This equation is derived from empirical data and theory for real-world circumstances.

Cautions and Useful Range of the Rule

This rule is heavily dependent on operation within the atmosphere, so it also depends on atmospherics and does not include effects on range of other attributes and parameters such as integration times, $1/f$ noise, clutter, and so forth.

Usefulness of the Rule

The rule can be used to

- quickly estimate the range performance impact of adding or deleting detectors
- compare range performance of similar FLIRs/cameras that have different numbers of detectors on their focal planes
- illustrate (and reiterate) one of the great effects of the atmosphere, and how difficult it often is to increase range by making simple design changes such as changing the number of detectors

Notes and Explanation

Classic radiometry indicates that the range of a FLIR is proportional to the fourth root of the number of detectors. However, FLIRs operate in the atmosphere. There is evidence that, when the atmospheric effects are considered, the range actually increases as the logarithm of the number of detectors.

This departure from classic radiometry illustrates the importance of the atmosphere's contribution to FLIR performance. Ranges should never be quoted without a statement about the assumed atmospheric conditions.

Sources

M. Kruer, D. Scribner, and J. Killiany. March 1987. Infrared focal plane array technology development for navy applications. *Optical Engineering* 26(3), 182–190.
J. Miller. 1994. *Principles of Infrared Technology*. New York: Van Nostrand Reinhold, 407–408.

RESOLUTION REQUIREMENT

Subject: ATP, spatial resolution, systems, target phenomenology

The Rule

The required system spatial frequency in cycles per milliradian can be estimated from:

$$F_r \approx \frac{1.5\ NR}{w}$$

where F_r = required system resolution stated in spatial frequency in cycles per milliradian

N = required number of cycles or line pairs or pixel pairs across the target

w = smallest dimension of a target in meters

R = range in kilometers

Basis for the Rule

The rule is based on empirical data and geometry.

Cautions and Useful Range of the Rule

The rule assumes that the resolvable spatial frequency is 65 percent of the system's cutoff frequency.

It assumes that one-dimensional accuracy is sufficient for this estimate

One should use the maximum slant range.

Usefulness of the Rule

It is useful for estimating resolution requirements, and it corresponds to several performance models' inputs and outputs.

Notes and Explanation

The required resolutions for an electro-optical system supporting a human operator depend on the target recognition (or detection) range and the number of pixel pairs or line pairs required for a human to perform the function at a given level. In general, humans can resolve frequencies in the range of 60 to 80 percent of the system's resolution. Also see "Target Resolution vs. Line Pair Rules" elsewhere in this chapter.

Incidentally, the Night Vision Laboratory has a slightly different version of this rule. They do not use the 1.5 multiplication factor.

Example

As mentioned previously in this chapter, a system that can resolve about five line pairs across a target can provide effective target recognition. Using this rule, we obtain

$$F_r \approx \frac{1.5\ NR}{w} = \frac{(1.5)\ (5)\ (1)}{3} = 2.5$$

Therefore, for a 3-m target at a range of 1 km, the required number of cycles per milliradian for recognition is about 2.5.

Source

B. Tsou. 1993. Systems design considerations for a visually coupled system. In *Emerging Systems and Technologies*, vol. 8, ed S. Robinson, of *The Infrared and Electro-Optical Systems Handbook*, executive ed. J. Accetta and D. Shumaker, Ann Arbor: ERIM and Bellingham Wa: SPIE., 520–521.

Backgrounds

A typical problem with earth-viewing surveillance sensors is that backgrounds can be complex and can include influences that reduce the chances of detecting targets. In general, it is desirable for any sensor to possess the capability of detecting targets in a variety of environments. The environments (or backgrounds) that the target is viewed against are often as bright and sometimes brighter than the targets themselves. This complicates and sometimes prevents the task of detection. To detect targets, the sensed background noise (or clutter) levels must be lower than the sensed target levels and low enough to provide acceptable false detection rates. In addition, the background may include spatial variation that includes a size distribution that matches the size of the target, although spatial variation in the background at all spatial frequencies will have a negative impact on the ability of a sensor to find the target, as we shall illustrate below.

The spatial variation in the scene is usually referred to as *clutter*. Clutter is a structured background phenomenon not spatially or temporally constant, and not independent of other noise sources. It is not a process that allows the signals to be combined in an RSS fashion; rather, they must be simply added to other noise sources. Clutter cannot be simply filtered out via simple signal processing like a DC background; it requires some more sophisticated analysis such as image processing. Again, the size has something (but not everything) to do with it, as it would be trivial to filter.

The physical sources of clutter for the chosen bandpass include large weather fronts, sun glints, clouds and cloud edges, variations in water content, lakes, and certain bright ground sources. For viewing from space, the altitude of the clutter sources will depend on the wavelength of operation. Only at wavelengths which penetrate the atmosphere will ground source be present for a space based sensor.

Examples of how clutter prevents detection are easily observed. A poppy seed dropped on a white piece of paper is easy to find. The same object dropped on a pattern of white and black dots is almost impossible to detect. In addition to the effect induced by clutter, detection systems must also deal with a background effect that is analogous to trying to detect the seed on a gray piece of paper.

Spatial, spectral, morphological, and temporal bandpass filtering are commonly used techniques in current sensor systems to reduce the effects of unwanted backgrounds. Spectral bandpass filters can reduce the resulting DC background level and restrict the light entering the sensor to a target signature that includes reduced amounts of clutter. Spatial filters can be used to reduce or attenuate the spatial frequency components of clutter that do not match the approximate target size. Figure 4.1. illustrates slightly different sized targets for which the spatial bandpass must be designed not to attenuate the targets. Any clutter that is also present in the bandpass will not be attenuated, either. The smaller targets are still difficult to identify in this picture. The clutter that gets through the bandpass is called *clutter leakage*. Clutter leakage is the artifact that results in false detections, which are generally handed to the tracking system.

This chapter includes a number of rules that deal with various sources of background radiation. While the rules are not all related to clutter and its effects, it remains one of the most difficult problems encountered in the design of EO tracking systems.

The environment of the target and the characteristics of the sensor system determine the ability of the sensor system to detect targets of interest. An ideal sensor would be able to measure the signature of a target without any noise contributions

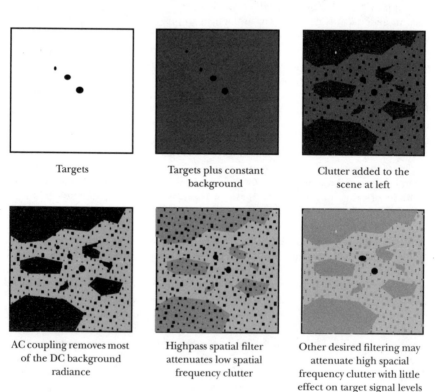

Targets	Targets plus constant background	Clutter added to the scene at left
AC coupling removes most of the DC background radiance	Highpass spatial filter attenuates low spatial frequency clutter	Other desired filtering may attenuate high spacial frequency clutter with little effect on target signal levels

Figure 4.1 Effects of Clutter and Filtering

from the sensor itself. However, ideal sensors do not exist. In addition to the sensor noise, there may be clutter present in the scene being viewed by the sensor. This scenario occurs in space-based sensors viewing targets with the earth as the background. In fact, sensor technology is sufficiently advanced that typical sensors viewing a scene with earth background will be background clutter limited rather than sensor noise limited. In the case of non-boosting targets, the clutter level in an earth background scene is likely to be so large that the target may not be detectable until well resolved.

GENERAL CLUTTER BEHAVIOR

Subject: Backgrounds, clutter, ATP

The Rules

Ben-Yosef[1-5] made several key observations about clutter and clutter power spectral densities in numerous papers. They are summarized in the following rules:

1. The dominant factors are variations in heat capacity, thermal conductivity, and solar radiation absorption of the terrain, along with solar reflectance.

2. In the presence of solar heating, variations in scene emissivity are not important, because the emissivity of natural ground materials may be greater than 0.9.

3. As solar insolation increases, the standard deviation of the irradiated power increases, and correlation length decreases, leading to clutter with higher spatial frequencies.

4. Terrain features that are unobservable in steady-state conditions may become apparent during warming or cooling transients.

5. The measured statistics of an IR scene are dependent on the spatial resolution of the sensor acquiring the data.

6. Although scene statistics from a high-resolution sensor may not be Gaussian, averaging by a low-resolution sensor will tend to produce Gaussian statistics.

Basis for the Rules

The rules are based on empirical observations and radiation and atmospheric theory; most of these observations are based on desert terrain. Desert terrain generally involves a "high" clutter condition, but urban clutter frequently is worse.

Cautions and Useful Range of the Rules

This rule is useful in infrared bands but does not necessarily apply to UV, visible, and radar bands. The background clutter must be dominated by natural terrain, which is not necessarily applicable to conditions of redout, urban environments, or other types of non-natural backgrounds.

Usefulness of the Rules

This rule is useful when trying to estimate the impacts of clutter for system design.

Notes and Explanation

Ben-Yosef and colleagues have published several papers analyzing clutter statistics. They developed a heat balance equation and experimentally validated it. Complete listings of the relevant papers can be found in the *IR & EO Handbook*.

References

1. N. Ben-Yosef, K. Wilner, and M. Abitbol. 1986. Radiance statistics vs. ground resolution in infrared images of natural terrain. *Applied Optics* 26, 2648–2649.
2. N. Ben-Yosef, et al. 1987. Temporal prediction of infrared images of ground terrain. *Applied Optics* 26, 2128–2130.
3. N. Ben-Yosef, K. Wilner, and M. Abitbol. 1987. Prediction of temporal changes of natural terrain in infrared images. *Proc. of SPIE*, vol. 807, 58–59.

4. N. Ben-Yosef, K. Wilner, and M. Abitbol. 1987. Natural terrain in the infrared: Measurements and modeling. *Proc. of SPIE*, vol. 819, 66–71.
5. N. Ben-Yosef, K. Wilner, and M. Abitbol. 1988. Measurement and modeling of natural desert terrain in the infrared. *Optical Engineering* 27, 928–932.

Additional Source

J. Lloyd. 1993. Fundamentals of electro-optical imaging systems analysis. In *Active Electro-Optical Systems*, vol. 4, ed. M. Dudzik, of *The Infrared and Electro-Optical Systems Handbook*, executive ed. J. Accetta and D. Shumaker, Ann Arbor, MI: ERIM, and Bellingham, WA: SPIE, 34–35.

REFLECTIVITY OF A SURFACE

Subject: Backgrounds, phenomenology

The Rule

In the visible wavelength range, the difference between the reflection of a wet surface and that of a dry one can be approximated by using the following equation:

$$\rho_w \approx \frac{0.9\rho_d}{(1.77 - 0.85\rho_d)}$$

where ρ_w = reflectance when the surface is wet
ρ_d = reflectance when the surface is dry

Basis for the Rule

This rule is a specific application of the more general expression shown below.

Cautions and Useful Range of the Rule

- It is valid for the visible bandpass.
- It assumes the wetting agent is water.
- It is an approximation only.

Usefulness of the Rule

The rule is useful for estimating the relationship of reflectances between wet and dry surfaces.

Notes and Explanation

For a thin layer of water in the visible spectral range, the index of refraction of water can be assumed to be 1.33; the hemispheric reflectance is about 0.08 with a transmittance of the liquid layer of 1.0.

The more complete rule states that

$$r(wet) = \frac{(1-\rho')\,\tau^2\,(1-r)\,(\rho'_d)}{n^2 - r_d\,\tau^2\,(n^2-1+\rho)}\left[1 - \left(\frac{n-1}{n+1}\right)^2\right]$$

where $r(wet)$ = the wet stack bulk reflectance
ρ' = hemispheric reflectance of the water for a Lambertian source
τ = transmittance through the liquid
ρ_d = hemispheric reflectance of the dry surface
n = index of refraction of the dielectric

Sources

W. Wolfe and G. Zissis. 1978. *The Infrared Handbook.* Ann Arbor, MI: ERIM, 3-11–3-12.

D. Kryskowski and G. Suits. 1993. Natural sources. In *Sources of Radiation*, vol. 1, ed. G. Zissis, of *The Infrared and Electro-Optical Systems Handbook*, executive ed. J. Accetta and D. Shumaker, Ann Arbor, MI: ERIM, and Bellingham, WA: SPIE, 144.

THE EARTH'S EMISSION AND REFLECTION

Subject: Backgrounds, viewing Earth from space

The Rule

There is a big difference in the brightness of the Earth between day and night when viewed from space in the visible and UV bands, but almost no difference for wavelengths beyond about 4 or 5 microns.

Basis for the Rule

Beyond approximately 3.5 microns, the component of total radiation and reflection from the Earth that is due to reflected sunlight is quite small (e.g., 20 percent of the total) because the emission from the sun is smaller by orders of magnitude as compared with its peak emission (which is near 0.5 μm, therefore the 280 K blackbody radiation tends to dominate). In addition, the temperature of the Earth's emitting surface does not change much within a day or so, even during the night.

Cautions and Useful Range of the Rule

The exact wavelength shown above can be adjusted somewhat, particularly if atmospheric effects are taken into account. However, this rule is quite general and can be used with confidence.

Usefulness of the Rule

This rule reminds us that the Earth background that a sensor encounters need not be considered as day or night missions are considered.

Notes and Explanation

The relative contributions of the day and night backgrounds of the Earth are of concern in dealing with clutter in the scene. Using this rule, we find that the processing system need not be concerned with the time of day for systems operating above 3.5 microns. However, in wavelengths shorter than this, the background will be distinctly different in day versus night cases.

CLUTTER PSD FORM

Subject: Backgrounds, clutter

The Rule

The clutter power spectral densities (PSDs) usually follow the form:

$$PSD(f) = C\left(\frac{1}{f^n}\right)$$

where $PSD(f)$ = power spectral density of the clutter as a function of spatial frequency. Usually, it has the units of in-band (watts/cm^2/sr/μm) per cycles/km in one dimension.

C = a constant, usually in the range of 10^{-7} to 10^{-2} for a PSD in watts

f = spatial frequency (how large the items causing the clutter are), usually in cycles per kilometer

n = a constant, usually between 1 and 3. The choice of the constant depends on the background conditions and the range of spatial frequencies your sensor detects on the background.

Basis for the Rule

This rule is based on curve-fitting empirical data.

Cautions and Useful Range of the Rule

Use with caution—this is a great generalization for most cases.

One-dimensional PSDs do not reflect the nonisotropic structure of the real three-dimensional world. Procedures for estimating two-dimensional PSDs from one dimension can be found in the *Infrared and Electro Optical Systems Handbook,* which is referenced in various sections of this book. Usually, the procedures entail incrementing the power (above constant) by +1.

Atmospheric attenuation of clutter should be considered. Absorption across the bandpass may be included in the PSD, but usually it is not. The user should estimate a range to the clutter and estimate the atmospheric attenuation across the expected path length.

Clutter is very bandpass sensitive, regardless of the "per micron" in the units. Using a PSD from an SWIR band to estimate LWIR clutter, or from a UV band to estimate visible clutter, is rarely accurate.

Usefulness of the Rule

This rule is useful for quick estimations of clutter in the absence of real data, drawing curves through one or two points, estimating the clutter rejection capability needed, and estimating the effects of pixel field of view on clutter induced noise.

Notes and Explanation

The power spectral density function is frequently used to describe background clutter. These can be one or two dimensional. For the one-dimensional models, the slope of the clutter PSD usually falls as $1/f^2$ to $1/f^3$ at lower frequencies (for clutter-causing phenomena that subtend a large angle), to $1/f^1$ to $1/f^{1.5}$ at higher spatial frequencies.

SIGNAL-TO-CLUTTER RATIO AS A FUNCTION OF FOV

Subject: Background, clutter, systems, point source detection

The Rule

Spencer[1] derived the signal-to-clutter ratio for unresolved targets as inversely related to the resolution, range to clutter, and characteristics of the clutter such that:

$$SCR \propto \frac{1}{(R \; IFOV)^{0.5n+1}}$$

where SCR = signal-to-clutter ratio
R = the range to the clutter
IFOV = the linear (radian not steradian) field of view
n = a constant describing the slope of the clutter's power spectral density (PSD). For real-world cases, this is usually approximately 3, but for odd backgrounds it may range from 0.5 to 6.

Basis for the Rule

This rule is based on two-dimensional PSDs and Fourier transform analysis.

Cautions and Useful Range of the Rule

The above assumes that the clutter and target are at the same range.

It does not consider atmospherics, which may affect the target differently from the clutter (e.g., if they are at different ranges, or if the target is out of the atmosphere but the clutter is not).

This rule assumes that the target is a point source.

This rule is uncertain whether the slope of the PSD is changing (e.g., at the break frequency) over the size of the spatial fitter (the integration limits).

This rule assumes square pixels on the FOV; rectangular or other odd-shaped pixels or optical distortion will have an effect on this rule, analysis, and conclusion.

Usefulness of the Rule

The rule estimates the effects of different clutter levels and different resolutions and/or ranges.

It underscores a basic effect on system performance from clutter.

For a given signal-to-clutter ratio, the range IFOV product is largely independent of the shape of clutter PSD. Thus, much of the fuss about PSD shapes is much ado about nothing.

Notes and Explanation

The mathematical derivation is as follows.

Clutter is usually expressed as

$$S(f) \approx \frac{K}{(f_x^2 + f_y^2)^{n/2}}$$

where $S(f)$ = clutter noise

K = a constant used to describe the power level and/or adjust the units

f_x = spatial frequency in cycles/length (usually cycles per kilometer) in the x direction

f_y = spatial frequency in cycles/length in the y direction

n = constant that determines the slope of $S(f)$

Now, by Fourier transform techniques, the variance in the clutter is

$$\sigma^2 = \int_{f_l f_l}^{f_h f_h} S(f_x f_y) \, df_x df_y$$

where σ^2 = the variance (recall that σ is the standard deviation)

f_l = the lower spatial limit of the bandpass that the sensor/image processor is using, usually assumed to be the low spatial frequency. This is usually equal to $1/NFP$ where N is the number of the pixels that the spatial filter is using and FP is the pixel footprint on the clutter. For example $N = 3$ times the pixel footprint for a 3×3 pixel spatial filter.

f_h = the higher spatial limit of the bandpass that the sensor/image processor is using, usually assumed to be the spatial frequency equal to one pixel footprint on the background

This integral is too difficult, so let us assume a relatively narrow power bandwidth and use the average to evaluate the integral. It turns out that a more rigorous treatment of the integral results in the same answer.

The bandpass of the operation is

$$\Delta F = f_h - f_l = \frac{1}{(FP)} - \frac{1}{N(FP)} = \left(\frac{N-1}{N}\right)\frac{1}{FP}$$

where ΔF = bandpass of operation

FP = the pixel's one-dimensional footprint on the clutter

N = number of pixels in the spatial filter (one dimension)

Substituting an average bandpass (accurate for narrow bandpasses), the narrow bandpass is about

$$\sigma^2 \approx S(f_x, f_y) \, \Delta f_x f_y$$

Substituting these into the form of $S(f)$, we obtain approximately $K(FP)^{n-2}$. Therefore, the standard deviation is the square root, or

$$\sigma \approx K^{1/2} (FP)^{[n/2] - 1}$$

The noise power on the detector from clutter is directly proportional to the standard deviation (not root-summed squared), so it can be expressed as:

$$P_{dc} = \sigma (A_o / R^2) (FP)^2$$

where P_{dc} = power on the detector from clutter
A_o = area of the optics
R = range to the clutter

Substituting yields

$$P_{dc} = K^{1/2} (A_o/R^2) (FP)^{0.5n+1}$$

Additionally, the footprint (FP) is equal to the range multiplied by the pixel's instantaneous field of view (IFOV), thus

$$P_{dc} = K^{1/2} A_o (IFOV)^{0.5n+1}$$

The power on the detector from the target is proportional to

$$P_{dt} \propto A_o/R^2$$

where P_{dt} = power on the detector from the target

Therefore, the SCR = P_{dt}/P_{dc}, or

$$SCR = \frac{(A_o/R^2)}{[A_o R^{0.5n-1} (IFOV)^{0.5n+1}]}$$

In this case the apertures cancel and, if the background and target ranges are identical, the above rule results.

If a specific noise spectrum is modified by a signal processor and sensor parameters, the local contrast can be determined by finding the difference in the radiated power between the center pixel and a window (usually a 3×3 to a 7×7) around it. Working through the mathematics, one finds that the clutter power is a relationship of the IFOV and range. The value of "n" tends to be near 3, so the effects of clutter diminish by the square to the third (5/2 being the preferred mathematically nominal value) of the pixel field of view.

If the PSD is flat, then it represents white noise and has an $n = 0$. For this case, the signal-to-clutter ratio will vary inversely with respect to range and resolution as

$$\approx K \frac{1}{R(IFOV)}$$

For the more commonplace case where $n = 3$, the signal-to-clutter ratio will vary as

$$\approx K \frac{1}{R(IFOV)^{5/2}}$$

The surprising feature is that the range and one-dimensional resolution are at the same power, regardless of the shape of the curve. In other words, the improvement in detection range that you will get by changing the IFOV is independent of the shape of the clutter PSD (when the target and clutter are at the same range)!

Source

This rule provided by Dr. George Spencer, 1995.

EFFECTIVE SKY TEMPERATURE

Subject: Backgrounds

The Rule

The temperature of the sky at zenith is estimated by:

$$T_{sky} = [0.7 + 5.95 \times 10^{-5} \delta \exp (1500/T_{air})]^{1/4} T_{air}$$

where T_{sky} = effective blackbody broadband temperature that would give a radiant
emission similar to that of the zenith sky
T_{air} = ambient temperature of the air
δ = water vapor pressure in millibars

This can be further reduced to multiplying the fourth root of the emissivity by the ambient air temperature to estimate the sky temperature in the 8- to 12-μm bandpass, or

$$(T_{sky} = \varepsilon^{1/4} T_{air}), \text{ which equals } [A + B\delta \exp (C/T_{air})]^{1/4} T_{air}$$

where ε = emissivity. The reader interested in a rather lengthy discussion of methods for estimating the effective emissivity of the cloudless sky should refer to Idso.[1]
A = bandpass-dependent coefficient (use 0.24 for 8 to 14 μm)
B = another bandpass-dependent coefficient (use 2.98×10^{-8} for the 8- to 14-μm band)
C = another bandpass-dependent coefficient (use 3000 for the 8- to 14-μm band)

For 10.5- to 12.5-μm bands, use $A = 0.1$, $B = 3.53 \times 10^{-8}$ and $C = 3000$.

Basis for the Rule

The rule is based on empirical curve fitting.

Cautions and Useful Range of the Rule

It is calculated for zenith only; one must adjust for other angles.

The above coefficients are solved for 10.5 to 12.5 μm and 8 to 14 μm only.

Usefulness of the Rule

The rule is used for a first-cut quick estimate or comparison of expected sky temperatures for background calculations for astronomy, air defense, satellite tracking, and so forth.

Notes and Explanation

When pointing up, a camera views through the atmosphere to outer space. In the absence of any bright object in the field of view, the radiant emission sensed by the device is equal to the 3 K effective temperature of the universe plus a higher temperature contribution from the intervening atmosphere. Idso[2] treated the atmosphere as a graybody at ground-level air temperature with an emissivity that was dependent on temperature and humidity.

Reference

1. S. Idso. 1981. A set of equations for full spectrum and 8- to 14-μm and 10.5- to 12.5-μm thermal radiation from cloudless skies. *Water Resources Research* 17(2), 295–304.
2. S. Idso and R. Jackson. 1969. Thermal radiation from the atmosphere. *Journal of Geophysical Research* 74, 5397.

Additional Source

D. Wilmot et al. 1993. Warning systems. In *Countermeasure Systems*, vol. 7, ed. D. Pollock, of *The Infrared and Electro-Optical Systems Handbook*, executive ed. J. Accetta and D. Shumaker, Ann Arbor, MI: ERIM, and Bellingham, WA: SPIE, 49.

SKY RADIANCE

Subject: Backgrounds, atmospherics

The Rule

For a heavily overcast sky, the distribution of radiance with zenith angle is a cardioid (a figure of revolution) as follows

$$L(\theta) = \frac{3\,E_d(0)\,(1 + 2\cos\theta)}{7\pi}$$

where $L(\theta)$ = radiance

$E_d(0)$ = downwelling irradiance on a horizontal surface in watts/meter^2 (Irradiance is received energy, expressed in watts/meter^2. It can be measured by pointing the instrument at the zenith.)

θ = zenith angle

Basis for the Rule

This is an empirical observation.

Cautions and Useful Range of the Rule

The sky coverage by clouds is rarely uniform. Therefore, the approximations of the type shown above, which assumes a uniform cloud cover and density, must be used with care.

Usefulness of the Rule

This type of rule allows the designer to estimate the change in the background that will be seen by a sensor as it views different parts of the sky.

Notes and Explanation

The sky illumination of the sea or land surface of the earth can be a significant factor in the upwelling (reflected) radiation that is observed. This rule makes it possible to estimate the contribution from the surface under heavy cloud conditions.

Source

J. Apel. 1987. *Principles of Ocean Physics.* Orlando, FL: Academic Press, 525.

Chapter

5

Cryogenics

Cryogenic engineering is important throughout the realm of electro-optics, as detectors require cooling for high sensitivity—especially in the infrared. Therefore, it can be said that cryocooling is a key enabling technology for modern sensors. Many systems require some form of refrigeration to very low temperatures, at least for some components. Again, this is especially critical to the sensitive infrared sensors that supply the largest market pull for the cryocooling market.

Usually, this cooling is accomplished by employing one of several methods including

- a liquid reservoir of cryogen (typically, for temperatures of less than 100 K)
- a mechanical refrigerator (for temperatures from 4 to 250 K)
- a Joule-Thomson blow-down expander (typically, for temperatures from 50 to 110 K)
- a thermoelectric cooler (for temperatures above 170 K)

This cooling power is usually delivered to the focal plane, the cold shield, the surrounding surfaces (e.g., the FPA mux and carrier), and sometimes the optics. Typically, an infrared detector focal plane array will require cooling, as well as its immediate surroundings. All of these parts are contained (the only exception being some space sensors) in a super-thermos bottle called a "dewar." For high sensitivity at long wavelengths, cooling of portions of the telescope (structure and optics) is also required.

Cryogenic cooling has traditionally been the bane of the electro-optical system engineer. Providing cooling by means of a perishable liquid always taxes the user's supply lines. In many applications, such as space, cryogen depletion limits the life of the sensor (e.g., IRAS and ISO). Employing a Joule-Thomson expander limits the operation time, usually to a few minutes. Traditionally, mechanical coolers were bulky, inefficient power consumers of low reliability. Thermoelectric coolers had poor efficiency, limited cooling, and poor performance in high shock/vibration environments.

For rapid, highly reliable cooling for a limited period, Joule-Thomson expander systems provide higher efficiency and reliability, but the lowest operation time (perfect for missiles, but horrible for cameras). Expanding a non-ideal gas through an orifice for cooling was first investigated by J. Joule and William Thomson (later to be named Lord Kelvin) in the 1850s. The effect was later refined and developed (near the turn of the century) by Linde and Hampton. Incidentally, Linde later went on to form a successful cryogen company. Another major development in the late 1970s and early 1980s used photolithography to define the gas flow passages and expansion in a thin disk. Today, modern Joule-Thomson expansion systems are compact, lightweight, and reliable.

The former Soviet Union concentrated on solid-state thermoelectric coolers in the 1950s and 1960s. Low-cost, high-reliability, limited-capacity coolers (with no moving parts) based on the Peltier effect were produced by the former Soviet Union and some American counterparts. Recent material advancements coming to light from the former Soviet Union, along with the promise of "thermogenic" materials, may allow these coolers to challenge the mechanical and Joule-Thomson coolers in the next century.

Mechanical cryogenic science is a rather new field, with very little research prior to World War II. Nevertheless, most cryocoolers in use today employ a cycle discovered by Rev. Robert Stirling in 1816. Since the 1940s, advancements in cryocooling have largely been in the incremental development of better dewars and more reliable coolers. The British concentrated their designs around providing a high-pressure source for (liquid) air expanders, while the Americans concentrated on mechanical cryocoolers. American military systems frequently employ Stirling- type mechanical coolers and (far less often) Vuilleumier or Gifford types.

Traditionally, employing a mechanical refrigerator resulted in limited system reliability and additional cost, weight, and power consumption. Mechanical cooler life of 5 hours was great in the early 1970s, 500 to 1000 hours was par for the 1980s, and now 4000 hours is common. Recent advancements in cooler technology have altered this archaic view, and the hardware advancements have migrated from the lab to commercially available products (e.g., the Magnavox 8000). Many of these modern coolers have *mean time to failures* (MTTFs) approaching other nonmechanical components (such as electronic circuit boards), use little power, and are substantially smaller than the beer cans at a successful critical design review party. Spurred on by military infrared applications, there are now numerous companies in several nations whose entire business is centered around the production of mechanical cryocoolers.

Significant advancement in mechanical coolers occurred in the 1980s with the development of the Oxford Cryocooler (employing ultra-pure working gases, clearance seals, and linear voice coil actuators). Another improvement was the pulse tube, which replaces the traditional piston in the expander with a slug of gas. The slug of gas acts like a mechanical piston, with slightly less efficiency and wear but higher reliability. Several companies are pursuing a pulse tube cryocooler to achieve 20,000 hour MTTF for tactical cryocoolers. Both of these advancements are expected to be integrated into several systems in the 1990s.

The future is bright, now that failure mechanisms are well understood for cryocoolers. When the reliability of small, production coolers approaches that of electronic components, the system engineer will find little benefit in employing electro-optical techniques that do not employ cryocoolers. This is because system reliability will be dominated by electronic failures rather than cooler failures. At the same time, recent detector advancements promise visible and infrared focal planes that do not require cryocooling.

For the reader interested in more details, their are only a few books that provide useful insight into to cryocooling. Walker's *Miniature Refrigerators for Cryogenic Sensors and Cold Electronics* is certainly one worth purchasing. Additional specific chapters on modern cryocoolers can be found in the *Infrared and Electro-optical Systems Handbook* and Miller's *Principles of Infrared Technology*. Occasionally, papers can be found in SPIE conferences and thermodynamic publications. Generally, every other year, there is an international conference of cryocooling technology that presents a wealth of state-of-the-art information.

COMPARISON OF COOLING TECHNOLOGIES

Subject: Cryocoolers, cooling, system selection of coolers

The Rule

For a given cooling application, Joule-Thomson coolers provide the most rapid cool-down with the lowest usage of electric power; thermoelectric coolers provide the slowest cool-down with the highest reliability and consume the most power. Stirling coolers have characteristics in between and are what you probably will use.

Basis for the Rule

This rule is based on the current state of the art.

Cautions and Useful Range of the Rule

Not all cooling situations allow the use of the three different types of coolers; often there is only one choice.

Remember, the state of the art changes; there are no absolutes in this rule.

Usefulness of the Rule

The rule is useful for understanding the top-level attributes of the various types of coolers.

Notes and Explanation

The four commonly used types of coolers are containment dewars, Joule-Thomson blow-down coolers, thermoelectric coolers (TEC) and Stirling cycle coolers. The type of cooler used depends on the system requirements, constraints, and specifications. Typically, containment dewars are used for laboratory or space systems that require cooling below 50 K or so. Both Stirlings and Joule-Thomson coolers are used in the temperature regime from 50 to 180 K. Usually, TECs are only used for detector temperatures above 180 K.

Expendable Joule-Thomson (J-T) cooling provides high reliability and rapid cool-down with no moving parts or consumption of power. However, they are limited by the amount of expendable gas that they carry. Usually, J-T cooling can be used only once. Therefore, they are often found on missile seekers or other short-lived applications. Some J-T systems recapture the expanded gas, repressurize, and are able to be used more than once. These systems typically are large, inefficient, and of low long-term reliability, so their applicability is usually limited to lab environments. The British have developed relatively small liquid air compressors that take in ambient air, dry it, purify it, and expel the nitrogen through a Joule-Thompson cryostat device. Outside of the UK, there have been few fielded applications of this, although this technique combines some of the best attributes of a J-T and mechanical system.

Thermoelectric coolers (TECs) provide high reliability, slow, low-efficiency cooling. TECs are usually limited to cooling a small detector by less than $120°$ C from ambient. Their electrical efficiencies are usually under 1 percent. They are normally found in sensors employing InGaAs, SWIR HgCdTe, lead salts, and uncooled arrays. Recent advancements from the former Soviet Union may broaden the applicability of this type of cooling.

Mechanical Stirling refrigerators seem to be the dominant technology for now and the near future. Stirlings provide thousands of hours of cooling with cool-down times of a few minutes and electrical efficiencies of 2 to 6 percent. These coolers weight a half a kilogram to a few kilograms (depending on cooling power) and are found in cameras and FLIRS.

JOULE-THOMPSON CLOGGING

Subject: Cryogenics, thermal design, Joule-Thompson coolers

The Rule

The clogging of a cryostat tends not to happen in conventional J-T cryostats when there is a concentration of less than about a two parts per million water (or CO_2) or a three parts per million concentration of a hydrocarbon, or particles of $6\,\mu m$ size.

Basis for the Rule

This is based on empirical observations of the state of the art.

Cautions and Useful Range of the Rule

The state of the art changes. Smaller orifices will clog more easily, larger ones will tend to clog less easily.

Usefulness of the Rule

The rule is useful for estimation of the cleanliness of gases needed for a J-T blow-down system.

Notes and Explanation

Joule-Thomson coolers function by passing a non-ideal gas through an orifice. The orifice and associated capillary tubes can clog which stops or reduces the flow of the gas. If contaminants exist in the gas bottle of a Joule-Thomson system in excess of the above, then clogging can be expected.

Additional Source

B. Bonney and R. Longsworth. 1991. Considerations in using Joule-Thompson coolers. *Proc. of the Sixth International Cryocoolers Conference.* Bethesda, MD: David Taylor Research Center, 231–44.

THE CRYOCOOLER SIZING RULE

Subject: Cryocoolers, systems

The Rule

The required cooling capacity for an application can be sized by either the steady-state heat balance or by the time allowed to cool a thermal mass to a cryogenic temperature, but not both; 10 percent should be added to the sizing criterion selected.

Basis for the Rule

This rule is founded on basic empirical observations, state-of-the-art technology, and common sense.

Cautions and Useful Range of the Rule

Always oversize the mechanical cooler capacity. The extra capacity will be needed sometime.

If performance at the end of life is important (or after some large number of cycles), it should be oversized by a large margin (perhaps 50 percent).

Usefulness of the Rule

This provides a criteria for margin on cryocoolers.

Notes and Explanation

When cooling a detector, more capacity is needed for a reasonable cool-down time. Application requirements are not reasonable wherein the steady state and cooldown requirements result in same refrigeration capacity. If this happens, reduce thermal mass or total heat load. When the cool-down and steady-state capacities are the same, the result is always an overcapacity at steady state and a detector that is too cold, or the need for a "demand" cooler with all kinds of electronics, or a cooldown time of hours instead of minutes.

Source

Rule provided by Mike Elias, 1995.

COOLER CAPACITY EQUATION

Subject: Cryocoolers, systems

The Rule

The available refrigeration capacity of a Stirling cryocooler can be estimated with an equation similar to:

$$Q_r = CPVFT_r$$

where Q_r = available refrigeration in watts

P = mean pressure of the working gas (or some other characteristic pressure)

V = swept volume in compression space

F = frequency of operation of the cryocooler (piston cycle frequency)

T_r = refrigeration temperature (temperature of the expansion space) in Kelvins

C = a constant (<1) to scale the equation and provide for the correct units

Basis for the Rule

The rule is based on the state of the art and curve-fitting empirical data.

Cautions and Useful Range of the Rule

The above relationship was developed based on large Stirling coolers (1 kW or more). It is not validated for small cryocoolers but can be used for them with caution. This rule is subject to the state of the art.

Usefulness of the Rule

It is valuable for quick estimations of the available refrigeration from a cryocooler and is also useful for estimations on what attribute of a cooler to change by what amount to achieve a different level of cooling.

Notes and Explanation

The available refrigeration (Q_r) is the amount of cooling in watts available at the cold finger, with all of the losses and inefficiencies included. This is the measure that sensor designers care about and is identical to the cooling capacity.

This equation suggests that the cooling capacity of a cryocooler varies in direct proportion to the speed at which it is run, the displacement of the compressor, and the pressure of the working gas and the temperature. However, the cooling capacity of small cryocoolers tends to be more closely related to the temperature at which the cooling is to be done. The cooling capacity of smaller cryocoolers also seems to be a function of the compressor case temperature of the cryocooler. This is because of the inefficiency of rejecting the heat and the inefficiency of the magnets in the motors.

Walker[1] suggests that C is on the order of 10^{-4}, for large Stirling coolers, when pressure is expressed in bars, frequency in Hz, and T_r in Kelvins.

Reference

G. Walker. 1989. *Miniature Refrigerators for Cryogenic Sensors and Cold Electronics.* Oxford, U.K.: Clarendon Press, 120–121, 158–160.

STIRLING COOLER EFFICIENCY

Subject: Cryocoolers, system power consumption

The Rule

The current state of the art for power efficiency for Stirling cryocoolers is about 30 to 40 W of input power for 1 W of cooling in the temperature region of liquid nitrogen.

Basis for the Rule

This rule is based on empirical observations of the sorry state of the art. This is about where current available cryocoolers are in terms of efficiency.

Cautions and Useful Range of the Rule

It is based on the state of the art in the mid-1990s; future developments may improve it slightly.

It is valid only between 70 and 80 K; the ratios are more efficient for cooling temperatures above 80 K and less efficient for cold end temperatures below 70 K.

This assumes the compressor case temperature is colder than about $40°$ C.

This assumes efficient (>85 percent) power converters and power supplies.

Usefulness of the Rule

The rule provides for a good estimate of the electrical power required to provide cooling, if a particular cooler has not been chosen.

Integral Stirlings (one-piece coolers with the expander as part of the compressor) are slightly more efficient (e.g., 20 percent).

Pulse-tube Stirlings (a variant of a Stirling that employs a "gas slug" in the expander instead of a mechanical piston) are slightly (e.g., 20 percent) less efficient.

Notes and Explanation

Classic Stirling cryocoolers operate by electric motors in the compressor to propagate a pressure change to drive a piston in the expander. The (wall plug) electrical power needed to drive the Stirling cycle is typically about 30 to 40 times the heat power to be removed. This is often called the *coefficient of performance* (COP). These electrical efficiencies are expected to progress to 20 to 25 W/W by the next century (for cold end temperatures in the 70 to 80 K range). If you need to reduce the cryocooler's power consumption, it is usually best to expend your efforts on reducing the dewar parasitics so that a smaller cooler can be used.

PARASITIC LOSSES IN DEWARS

Subject: Cryogenics, systems, FPAs

The Rule

Dominant cryogenic heat loads are usually the parasitic ones, not detector dissipation. Any calculation of heat loads must include radiation and conduction losses. This is true even with a photoconductive HgCdTe in a common module dewar.

Basis for the Rule

This rule is a generalized assertion backed by empirical observations of the state of the art.

Cautions and Useful Range of the Rule

Parasitic losses become more dominant as the temperature decreases.

The state of the art changes, so these losses are likely to become less dominant as better insulated dewars are developed.

May not be true for large visible CCD arrays with minimal cooling.

Also, this rule is becoming less valid as technology progresses. New large-scale dense focal plane arrays with signal processing incorporated into the multiplexers may drive the heat load more than parasitics.

Usefulness of the Rule

This rule is useful both for underscoring a serious caution and for estimating the dominant thermal loads.

Notes and Explanation

At cryogenic temperatures, the dissipation of the cold electronics usually fails to dominate the thermal loads. Usually, the parasitic connections to the warm world do. Any calculation of heat loads must include radiation and conduction losses. After hundreds to thousands of detector/dewars have been made, as with the US Army common modules (a series of standardized infrared detector/dewar packages), the dewar costs tend to dominate. This is largely because it is difficult to control the parasitic thermal loads at cryogenic temperatures.

SINE RULE OF IMPROVED PERFORMANCE FROM COLD SHIELDS

Subject: Cryogenics, radiometry, systems, detectors, optics

The Rule

The reduction in background noise (and therefore increase in sensitivity for a BLIP detector) realized by employing a cold shield can be estimated by

$$\frac{A}{[\sin(\theta/2)]}$$

where A = a cold-shield efficiency constant depending on the design and tolerances. Usually, the efficiency of a cold shield is between 90 and 95 percent, so A, would be between 0.9 and 0.95.

θ = full field-of-view angle (not half angle); in a 15° optical system, θ = 15°

Basis for the Rule

This rule is based on radiometry and empirical observations and approximations.

Cautions and Useful Range of the Rule

Nothing is perfect, a perfect cold shield does not exist, and a degradation constant
is required (hence the A, which is always less than one).

The above relationship is useful for all wavelengths. For small angles, the solid
angle may be used.

Usefulness of the Rule

This little rule provides easy estimation of the benefits of a cold shield over no cold
shield (see Fig. 5.2).

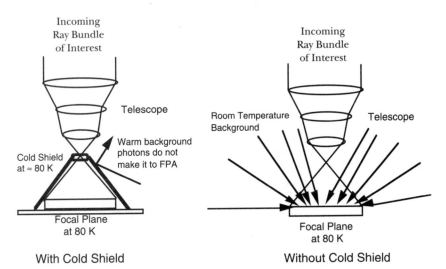

Figure 5.2 Performance with and without Cold Shield

Notes and Explanation

The rule is correct for the expected reduction in the noise associated with the background. The actual background flux is reduced by the square of the above figure.

A focal plane reacts to the background as well as the target signal. A flat detector with no shielding responds to energy from a background of a solid angle of 2π steradians—it "sees" everything. Due to projected angle (cosine effects), the actual approximate value of the flux is calculated by using just π steradians (see "Lambert's Law," page 297). However, the important point is that it only detects the useful target energy from a cone defined by the optics. It is a wise practice of the IR industry to limit the background that the detector "sees" by including a cold light shield. Ideally, this shield should limit the incoming radiation to the required field of view. If so, the benefit of including one is to reduce the background noise by $1/(\sin \theta/2)$. The benefit of cold shielding increases with narrow fields of view and longer wavelengths. In fact, the above rule also approximately equals two times the f# (ratio of the effective focal length to the effective aperture), because the f# defines the cone's solid angle.

However, a cold shield is never perfect. Manufacturing tolerances require that it be somewhat oversized, and "stray light" can leak in from its attachments and seams. Additionally, cold shield design is impacted by the available back focal length and thermal load. Thus, a 90 to 98 percent efficiency typically can be achieved, hence the "A" degradation factor. The system engineer usually puts some margin into the cold shield and, in the best cases, expects to get only between 90 and 95 percent effectiveness (A).

COLD SHIELD COATINGS

Subject: Cryogenics, radiometry, systems, dewar design, detectors, optics

The Rule

A cold shield should be black on the inside and shiny on the outside. That is, of high emissivity on the side viewed by the detector and of low emissivity on the side not viewed by the detector.

Basis for the Rule

This rule is based on radiometry, ray tracing, and empirical observations and approximations. The detector should not directly see a highly reflective surface, as it will reflect high background radiation onto the detector. However, to keep the thermal load low, the outside of the cold shield should be of low emissivity so that it does not absorb thermal radiation.

Cautions and Useful Range of the Rule

If the optics are contained within the cold portion of the system, and the self-radiation from the cold train is a concern, then in some peculiar designs it may make sense to have the cold shield of low emissivity in the inside. There is at least one production LWIR system that switched to shiny interiors of the cold shields after it was discovered that the expensive coating provided minimal increase in performance.

Some designs have been developed for highly reflective baffles and cold shields for special cases.

Nothing is perfect; a perfect cold shield does not exist. Therefore, this rule applies to the emissivity and reflectivity of the interior and exterior of common cold shields, and sometimes can be violated with an increase in performance.

Usefulness of the Rule

This is useful to underscore a basic tenet of dewar design and to remind one of the practical considerations for cold shield design. Usually, coating the inside black and the outside shiny takes several weeks.

Notes and Explanation

Initially, one might think that the cold shield should be of low emissivity (shiny). A review of Planck's equation certainly supports this, as a shiny cold shield interior will clearly radiate less to the FPA. However, this is one of the cases where pragmatic considerations usually overpower theoretical ones. If the cold shield is made shiny, it will reflect the warm surroundings onto the focal plane (as shown in Fig. 5.1) and reduce its overall effective efficiency. If it is black and Lambertian, only a small portion (typically 5 to 10 percent) of the warm photons will make their way to the FPA. If one can keep the unwanted photons leaking through the cold shield to 10 percent of the other noise photons, there will only be about a 5 percent increase in total noise for BLIP conditions (less for non-BLIP conditions).

Moreover, the apparent problem of the cold shield emission is usually reduced to negligible concern because usually the cold shield is far colder than it needs to be for radiometric purposes. It is usually thermally tied to the FPA and therefore cooled to a temperature close to that of the FPA. Radiometry shows that rarely will emission from the cold shield be a contributing factor.

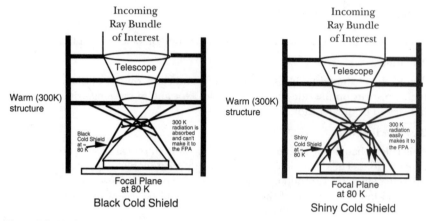

Figure 5.1 Black and Shiny Cold Shields

It is usually beneficial to make the exterior of the cold shield (the part the detector does not "see") as shiny as possible to reduce thermal load. This side is usually coated with gold or aluminum.

RADIANT INPUT FROM DEWARS

Subject: Cryogenics, systems, radiometry, FPAs

The Rule

Actual radiant background flux density from a dewar is often larger than that predicted by the Planck function. The difference is difficult to predict analytically, but it seems to be between 25 percent and a factor of two higher.

Basis for the Rule

This is based on empirical observations of existing systems. It probably results from less than expected cold shield efficiency and cold shield leaks.

Cautions and Useful Range of the Rule

Obviously this is a generalization, so it should only to be used a coarse guide.

This rule is sensitive to dewar geometry, cold shield design, manufacturing tolerances, and so forth.

Usefulness of the Rule

It is a reminder to add an extra factor to background flux calculations.

Notes and Explanation

Reflection of background irradiance, unforeseen cold shield leakage, and imperfect optical filters tends to cause an increase in the radiation impinging on a detector over what is routinely predicted by the Planck function. The effect is more pronounced for low-background LWIR detectors of wide natural response.

Source

J. Vincent. 1989. *Fundamentals of Infrared Detector Operation and Testing.* New York: John Wiley & Sons, 196–199.

THERMAL CONDUCTIVITY OF MULTILAYER INSULATION

Subject: Thermal insulation, cryogenics systems

The Rule The thermal conductivity of silk net/double-aluminized mylar multilayer insulation (MLI) can be estimated by:

$$k = (8.962 \times 10^{-4}) \, N^{1.56} \frac{T_h + T_c}{2} + \frac{5.403 \times 10^{-6} \varepsilon \, (T_h^{4.67} - T_c^{4.67})}{(T_h - T_c) \, N}$$

where k = thermal conductivity in μW/m K

N = layer density in layers per centimeter

T_h = hot temperature

T_c = cold temperature

ε = broadband room temperature emissivity of the blanket

Additionally, for Tissuglas/double-aluminized mylar, a different approximation is often used:

$$k \, [\mu\text{W/m K}] = \frac{(3.07 \times 10^7) \, (T_h^2 - T_c^2) - (2.129 \times 10^{-10}) \, (T_h^3 - T_c^3) \, N^{2.91}}{T_h - T_c}$$

Basis for the Rule

The rule is based on empirical observations derived from data in the references. Thermal conductivity has eluded a strict analytical approach due to difficulties in predicting the contact pressure, contact area, number of wrappings per cm, interstitial gas pressure, and material properties.

Cautions and Useful Range of the Rule

The constants are based on the state of the art, which changes.

The above rule assumes no cracks in the blanket. (Cracks can have thermal conductivity a factor of hundreds higher.)

This is approximately valid in the temperature range from 70 to 300 K.

Some measurements have constantly indicated that the thermal conductivity is higher than that predicted by the above equations, so the above should be used as a crude guide only.

Usefulness of the Rule

This rule is useful for quick estimations of heat loss through an MLI blanket and thereby estimating the thickness and density of an MLI blanket.

Notes and Explanation

Often, MLIs are employed in electro-optical sensor design. These blankets usually are implemented by wrapping the component (or sensor) many times in a thin MLI blanket like an Egyptian mummy. These were developed as a general-purpose insulator for space use but are also sometimes used in dewars.

Some investigations[1] suggest that the thermal properties of MLI blankets are optimized at about 30 layers per cm of thickness.

When employing MLI in an E-O sensor, two cautions must be noted. First, when handled, chips of metal sometimes flake off of the blanket and will always find their way to critical surfaces (e.g., onto optics and across electrical leads). Also, MLI wrappings greatly increase the surface area and can trap air pockets within the insulation layers. Therefore, MLI can greatly lengthen the vacuum pumping and any bake-out process.

Reference

1. P. Blotter and J. Batty. Thermal and mechanical design of cryogenic cooling systems. In *Electro-Optical Components*, vol. 3, ed. W. Rogatto, of *The Infrared and Electro-Optical Systems Handbook*, executive ed. J. Accetta and D. Shumaker, Ann Arbor, MI: ERIM, and Bellingham, WA: SPIE, 370–373.

J-T GAS BOTTLE WEIGHTS

Subject: Cryocoolers, systems, Joule-Thomson cryocoolers

The Rule

The weight of a containment bottle for the gases for a Joule-Thomson cryostat can be estimated from the following relationships:[1]

$$W_s \approx 0.000373 \, (V)^{0.86} \, (P)^{0.49} \, (FS)^{0.7}$$

$$W_c \approx (2 \times 10^{-8}) \, (V)^{1.067} \, (P)^{1.5} \, (FS)^{1.36}$$

and for high-tech composites:[2]

$$W_{HT} \approx \frac{PV}{1 \times 10^6}$$

where W_s = weight of a spherical container in pounds
W_c = weight of a cylindrical container in pounds
W_{ht} = approximate weight, in kilograms, of a high-tech composite container
V = volume of the container in cubic inches
P = operating pressure of the gas in pounds per square inch
FS = a safety factor (usually about 2)

Basis for the Rule

This is based on curve-fitting to empirical data backed by the state of the art.

Cautions and Useful Range of the Rule

Since this is based on state of the art, it will change with time and advances.

It does not include mounting, handling, valves, or line hardware.

It assumes normal J-T pressures from 3000 to 6000 lb/in².

Generally, this is accurate to within ±15 percent for volumes from 10,000 to 40,000 in³; it probably underestimates the weight of smaller bottles.

A cylinder's L/D ratios are not included in the equations.

Usefulness of the Rule

This is valuable for estimating gas bottle weight.

Notes and Explanation

Joule-Thomson cryostats require high-pressure gas to be blown through them to produce cooling. The significant contribution to weight and size is not the cryostat but the containment bottle for the high-pressure gas (usually nitrogen or argon).

Prominent technological advancements are occurring in tank production for lightweight propulsion systems. This same tank technology can be applied to the reservoir of high-pressure gas. For pressures required by J-T systems, the dry weight of a high-tech composite wound tank can be estimated by the third equation above. Advancements is high-pressure tank technology could elevate the constant in the denominator by a factor or two or three. This will lead to even lighter tanks in the future.

When calculating weight, be sure to add the weight of the gas. As any scuba diver knows, the weight of the gas is not trivial.

Example

If one requires a pressure of 10,000 lb/in^2 and a volume of 50 in^3 at room temperature to perform a given cooling function, the above relationships can be used to estimate the mass of the containment bottle. If spherical, it would weigh

$$W_s \approx 0.000373 \, (50)^{0.86} \, (10,000)^{0.49} \, (2)^{0.7}$$

or about 1.6 lb or 725 g. Additionally, a cylindrical bottle would weigh:

$$W_c \approx (2 \times 10^{-8}) \, (50)^{1.067} \, (10,000)^{1.5} \, (2)^{1.36}$$

or something on the order of 3.3 lb or 1500 g. Now, if weight is a critical concern, then one might want to employ high-tech composite fiber tanks and estimate the weight as:

$$W_{HT} \approx \frac{50 \times 10,000}{10^6}$$

or 500 g.

Additionally, the weight of the gas must be included. Assume that the gas is nitrogen. First, the number of moles must be estimated. To do this, we must make some conversions. The room temperature is 300 K, the volume is $(50 \times 2.54 \times 2.54 \times 2.54/1000)$ or 0.82 L, and the pressure is $(10,000/30)$ or 333 atmospheres. We can find the volume per mole from the familiar equation

$$\frac{RT}{P} \text{ or } \frac{0.082 \times 300}{330} = 0.075 \text{ L/mole}$$

and we have 0.82 L, so there are 11 moles of nitrogen. Eleven moles of nitrogen means that we have 11×28 or 308 g of gas (remember Avogardro's number, $[6.02 \times 10^{23}]$).

References

1. M. Donabedian. 1978. Cooling systems. Chapter 15, *Infrared Handbook*, ed W. Wolfe and G. Zissis. Ann Arbor, MI: ERIM, 15-14–15-16.
2. J. Miller. 1994. *Principles of Infrared Technology*. New York: Van Nostrand Reinhold, 204–206.

6

Detectors

Detectors are a quintessential component in any electro-optical system. They are most useful when combined into a dense two-dimensional array called a *focal plane array* (FPA). Traditionally, detector arrays have had the dubious position of being the system sensitivity limiter, resolution limiter, and cost driver. However, advances in the technology are making the FPA less of a performance and cost concern. This chapter includes rules relating to their performance, manufacture, and usefulness.

The conversion from light to electricity can occur via several mechanisms. The most popular are:

1. a *thermal* effect, where the light raises the temperature of some material (such as Ge or in silicon microbolometer arrays). This is also called a *bolometric effect*. This is seeing a resurgence in use with uncooled microbolometer arrays.

2. the *photoconductive* effect, where the resistance is changed so that the conductance of a material is altered with the level of irradiance. This requires a bias voltage and circuit to read the change in current through the detector.

3. the *photovoltaic* effect, where the light generates a voltage (or current) in a material. This requires a readout circuit that can sense this change in voltage (or current).

As can be summarized from the above mechanisms, detector physics and semiconductor physics are closely related. The advancements in one directly contribute to the advancements in the other.

The biologic eye is a wonderful detector and is covered in a separate chapter. To date, it holds the distinction of being the most sensitive and highest-resolution device that can be built in nine months by unskilled workers. Perhaps the first non-biological mechanical photonic detector was Hershell's blackened thermometer, which he used to discover infrared (IR) radiation. In the past few hundred years, detector science has made many promises and a few advancements.

The photoconductive effect was first reported by Willoughby Smith in 1873 while studying selenium crystals. Also, much advancement in film technology occurred in the latter half the 1800s, but silver halide does not respond well in the ultraviolet or beyond about 1.2 µm. Additionally, films do not lend themselves to the electrical digitization needed by modern processors and displays.

Albert Einstein won a Nobel prize in physics for explaining the photoelectric effect. He determined that a photon has a characteristic energy hv (Planck's constant times the frequency), and an energy greater than this is needed to free an electron in a photoelectric effect. Today, photoconductors, photovoltaics, and quantum well detectors use these quantum phenomena to respond to radiation of an energy greater than (or a wavelength shorter than) 1.24 divided by the photon's energy in electron-volts.

The desire for television drove the detector industry to develop vidicon uses for the visible in the 1940s, with refinement in the 1950s and 1960s. These are bulky, fragile, require high voltages, and are difficult to calibrate. However, they are producible, manufacturable, stable, and provide adequate resolution. The 1970s also saw the invention of the *charge coupled device* (CCD) and *charge injection device* (CID). The CCD can be easily implemented in silicon, providing direct conversion from photons to electrons (in the visible portion of the spectrum). Military and scientific users immediately employed the CCD, with commercial products finally becoming commonplace in the home video cameras of the 1990s. Currently, 5000× 5000 visible CCDs are available that match (or exceed) the resolution of tubes, but the "highly manufacturable at low cost" attribute of tubes probably will not be completely duplicated until the turn of the century for these high-resolution arrays.

Infrared and ultraviolet (UV) sensitive arrays are less mature. Ultraviolet arrays are often "enhanced" visible arrays with special antireflection coatings. Arrays of special materials (e.g.,TnO) and arrays of photoemissive channels have been developed for UV imaging.

During World War I, T. Case experimented and produced silicon and thallous sulfide (also called by *lead sulfide* or *galena*) detector cells. As a true father of the detector industry, Case created devices that were difficult to manufacture and had reliability problems. Then, World War II and the tension that preceded it spurred great investment by Germany and Great Britain, and some interest in the USA, in electro-optic sensor development. Lead sulfide continued to be refined, in the 1940s, by the Americans (e.g., Cashman, Case, and others) and in Germany by Edgar Kutzscher. Following the war, its development slowly progressed, with single-element systems and small arrays appearing in missiles in the 1960s. Today, arrays of up to 256 PbS elements are routinely manufactured and used in numerous military and commercial applications. However, it was the former Soviet Union that led some of the greatest advances in lead-salt detector technology and effectively implemented them into many high-tech systems.

In the 1960s and 1970s, many westerners were convinced that a specific mixture of mercury-telluride and cadmium telluride would make an excellent detector. Billions of research dollars flowed from Capitol Hill to military laboratories and then to American industry. Development in other materials waned, and today it is the standard infrared detector material being made into arrays as large as 480× 640, (with several American and European companies producing fine focal planes).

IR detectors are seeing some surprising advancements. Military research in the 1980s used micromachining to develop uncooled silicon bolometer arrays and uncooled ferreoelectric arrays, which are encroaching on the sensitivity of traditional materials. Additionally, molecular beam epitaxy allows us to custom form lattice structures yielding advancements in HgCdTe and custom quantum well detectors.

The 1990s have witnessed minor advancements in the science of producing uncooled arrays, InSb, and HgCdTe, and we also have seen a further price reduction and the development of UV and visible silicon detectors. Perhaps the military downturn of the 1990s will be responsible for a large cost reduction across the spectrum of detector materials as manufacturers seek new business in lower-cost markets. If so, this truly may be the hallmark of commercially available IR systems.

Generally, detectors are categorized by nature, depending upon their responses. Table 6.1 summarizes the typical wavelengths of commonly used detector materials.

Table 6.1 Typical Wavelengths of Commonly Used Detector Materials

Material	Typical Useful Spectral Region (μm)	Notes
CdS	0.3 to 0.55	Rarely used today.
GaAs	0.6 to 1.8	Linear arrays are commercially available. Doping with phosphorus extends the cutoff wavelength.
GaAs quantum well infrared photodiode (QWIP)	2 to 20	This material is tunable at time of manufacture, limited in spectral bandwidth, and very suitable to dense arrays and low-cost production
Ge:X	2 to $\approx$100	Doped Ge has long been an IR detector, with one to a few elements per array. Ge:Hg can respond as low as 2 μm, while Ge:Ga can respond at 100 μm at 3 K.
HgCdTe	2 to 22	This material is tunable at time of manufacture, arrays to 12 μm are commonly available
InGaAs	0.8 to 1.7	Cutoff can be extended to about 2.6 μm by adding phosphor. Usually, only thermoelectric cooling is required.
InSb	1 to 5.5	Some versions have response into the visible.
LiTaO$_3$	5 to 50	Pyroelectric materials with two-dimensional arrays are available.
PbS	1 to 3	Usually photoconductive, so two-dimensional arrays are rare, although many linear arrays are made.
PbSe	2 to 5	Two-dimensional arrays are rare, but many linear arrays are in production. New developments from the former Soviet Union may make this material a serious contender for MWIR applications
Pt:Si	1 to $\approx$5	Pt:Si is highly uniform and producible in large formats but offers low quantum efficiency at MWIR wavelengths.
Si	0.3 to $\approx$1.0	Red-enhanced and blue-enhanced versions are available; it lends itself to IC manufacture (monolithic CCDs).
Si:X	0.3 to 26	Doping silicon allows detection into the LWIR; requires cooling well below liquid nitrogen temperatures.
Si micro-bolometers	3 to 25	Via micromachining, silicon can be made into a tiny bolometer with a thermal response; this lends itself to dense arrays.

For the reader interested in more details or a more thorough understanding, numerous books and journals are available. As an introduction, we would suggest first checking out the appropriate chapters in some of the following:

- *Infrared Systems Engineering* (Hudson)
- *Thermal Imaging Systems* (Lloyd)
- *Electro Optical System Analysis* (Seyrafi)
- *Infrared Technology Fundamentals* (Spiro and Schlessinger)

Among the more modern texts, reflecting developments of the 1990s, are:

- *Electro-Optical Imaging System Performance* (Holst)
- *Principles of Infrared Technology* (Miller)

The following handbooks also have useful chapters:

- *Electro Optics Handbook* (Burle)
- *Infrared Handbook* and the *IR/EO Handbook* sets

For more detailed background of a scientific nature one should see the following:

- *Optical Radiation Detectors* (Dereniak and Crow)
- *Fundamentals of Infrared Detector Operation and Testing* (Vincent)
- *Infrared Detectors* (Willardson and Beer)

For the latest breaking technical developments, it should be noted that SPIE and IRIA/IRIS have regular detailed sessions on detectors. The journals that seem to cater to the scientific and detailed engineering needs of electro-optical engineers in this area are: *Infrared Physics and Technology, Optical Engineering,* and IEEE's *Transactions on Electron Devices.* Trade journals frequently provide valuable, up-to-date information on the technology that is of critical use to the anyone trying to select or use a FPA. Valuable trade journals in this discipline include *Photonics Spectra* and *Laser Focus,* which have a annual detector sections and frequent articles. Other sources of articles of a subtechnical nature on state-of-the art news include *Military and Aerospace Electronics, Defense Electronics,* and *Aviation Week and Space Technology.* Finally, an interested reader should review the corporate publications of those active in detector manufacture. These journals frequently have articles that provide an excellent marketing and technology review of a given company's products.

RISE TIME

Subject: Detectors, physics and engineering, electronics, optoelectronics

The Rule

The switching time characteristics of nonamplified junction photodetectors is determined by the RC performance down to rise times of 10 ns. In practice, Chappell gives the rise time as approximately:

$$T_r = 2.2\ RC$$

where T_r = rise time in seconds
R = resistance in ohms
C = capacitance in farads

Basis for the Rule

This is based on electrical engineering of photo-detectors.

The above equation above is based on math and filter theory, assuming a rise time between 10 and 90 percent of peak response. Accordingly,

$$T_r = T(90\ \text{percent of response}) - T(10\ \text{percent of response})$$

$$0.1 = 1 - e^{-T_1/(RC)}$$

$$0.9 = 1 - e^{-T_2/(RC)}$$

$$e^{-T/(RC)} = \frac{1}{1-0.1} \quad \text{and} \quad e^{-T_2/(RC)} = \frac{1}{1-0.9}$$

$$e^{(T_2-T_1)/RC} = \frac{1-0.1}{1-0.9}$$

$$T_r = T_2 - T_1 = RC\ \ln\!\left(\frac{1-0.1}{1-0.9}\right)$$

$$T_r = 2.197\ RC$$

Cautions and Useful Range of the Rule

The above rule depends on the detection mechanism, detector material and architecture, circuit implementation, and the state of the art.

This may not be valid for extremely short rise times, as other effects may contribute to the shape of the pulse.

Usefulness of the Rule

The rule is useful for estimating a detectors rise (or response) time for optical communications or scanning systems. Often, these systems result in very quick, effective integration times, so rise time can be critical. It is also useful for estimating the performance of photodetectors in optoelectronic applications.

Notes and Explanation

The response time of a photo-detector should be much less than the dwell or integration time. (This can be a concern for many materials, such as lead sulfide and bolometers.)

Lloyd[1] points out that many photo-detectors behave like a single RC lowpass filter in this respect, with the first equation dominating.

Commercially available silicon photodiodes have rise times in the neighborhood of 100 ns to 10 μs. However, Si-PIN photodiodes tend to be faster, with rise times closer to 10 ns, and avalanche photodiodes can be even faster, with rise times of about one-third of a nanosecond. Chappell[2] points out that the connected load resistance can be used for the resistance figure, and the total component capacitance can usually be found on a data sheet.

References

1. J. Lloyd. 1979. *Thermal Imaging Systems.* New York: Plenum Press, 108.
2. A. Chappell. 1978. *Optoelectronics Theory and Practice.* New York: McGraw Hill, 187.

Additional Source

Additional information supplied by Dr. Robert Martin, 1995.

SENSITIVITY TO TEMPERATURE

Subject: Detectors, operating temperatures, cryogenics

The Rule

Visible silicon detector arrays increase sensitivity by about 1 percent for each 1° C that they are cooled below ≈ 30° C. Lead salts (especially PbSe) increase sensitivity about 3 percent for each 1° C cooled below room temperature. MWIR HgCdTe increases sensitivity about 7 percent for each 1° C that it is cooled below ≈ 220 K. LWIR HgCdTe increases about 15 percent for each 1° C below ≈100 K.

Basis for the Rule

This rule is based on approximation of detector physics and empirical observations of the current state of the art. As the detectors temperature is increased, the internal noises rise, and the effects of offset and gain nonuniformity may become greater. For example, dark current will double for every ≈5 K increase in LWIR HgCdTe operating temperature above 80 K.

Cautions and Useful Range of the Rule

Sensitivity versus temperature is not linear; it is a curve, so do not overuse this rule. This rule is dependent on the state of the art (which changes) and assumes normal focal planes. Additionally, scaling should be limited to ± 20° about the normal operating temperature. Clearly, a temperature is reached for any material in which additional cooling provides no additional system-level sensitivity.

Usefulness of the Rule

It provides quick estimations of the amount of sensitivity that can be gained by cooling the detector a few more degrees.

Notes and Explanation

When a focal plane array (or an avalanche photodiode) is cooled, the noise decreases while quantum efficiency remains constant, causing its overall sensitivity to improve. Other benefits from additional cooling may be increased uniformity, longer wavelength response, and the ability to integrate longer. However, a temperature eventually will be reached at which further cooling provides minimal gains, as the total noise become dominated by photon shot noise and multiplexer readout noise (the "diminishing returns" concept). Additionally, multiplexer and bias circuitry may start to fail if operated at temperatures colder than they are designed to withstand.

With some detector materials (most notably, HgCdTe) the system designer can increase the FPA operating temperature to a point that the detector noise (Johnson, dark current, 1/f, and so on) becomes the dominant noise source, above photon shot noise. This will result in higher system reliability (due to increased cooler life), quicker cool-downs, and less dissipated heat. However, a caveat should be noted: the cutoff wavelength may change (as it does for HgCdTe).

This also applies to trade-offs of the cooling impact versus the impact of a more sensitive detector. It is possible that the minimum-cost sensor system is one that reduces the specification on the detector and cools it a few more degrees to maintain sensitivity.

PERFORMANCE DEPENDENCE ON RESISTANCE AREA PRODUCT (R_oA)

Subject: Detectors, semiconductor detector response

The Rule

The R_oA of a semiconductor material seems to change with temperature based on

$$R_oA = A10^{B/T}$$

where R_oA = resistance area product in ohm $\bullet$ cm^2
    A = a constant within the temperature range
    B = another constant within the temperature range
    T = temperature in Kelvins

Basis for the Rule

This is based on empirical observations of semiconductor detectors (e.g., HgCdTe and InSb).

Cautions and Useful Range of the Rule

This provides an estimation only.

One must make sure that A and B are correct for the material and temperature range.

Do not use the rule to compare different materials. For example, InSb changes more rapidly than HgCdTe in 80 to 100 K regime.

Usefulness of the Rule

The system engineer often wishes to know how the R_oA changes with temperature, since temperature is the only variable that can be controlled once the array is made. This equation allows easy scaling from one temperature to another.

Notes and Explanation

The above equation was determined by curve fitting published data on HgCdTe and InSb. The basic shape of the equation seems to hold well; however, the challenge is determining the constants. This is far more subjective and difficult, and it changes with the state of the art. The authors suggest that the interested reader scale from FPA manufacturer data or published data. Generally, R_oA for LWIR HgCdTe reduces by a factor of two for every $\approx$3 K increase, and MWIR HgCdTe's R_oA is halved for every 5 to 8 K increase. For photovoltaic LWIR HgCdTe at about 80 K, Martin[1] suggests the following handy relationship:

$$R_oA \approx (10^{[12 - \lambda \, (\text{cutoff})]})$$

Many semiconductor detectors also have other noise terms that vary in a similar fashion. Generally, dark current follows the same mathematical form except one should also consider the area of the detector for scaling dark current:

$$I_{d1} = I_{d2}\left(\frac{A_{d1}}{A_{d2}}\right)10^{K/T}$$

where I_{d1} = dark current in amperes (or nanoamperes) for a new detector, or the
     same detector at a different temperature

I_{d2} = dark current for the known detector in amperes or nanoamperes (if
     none are familiar, vendors can usually tell you what the dark current is
     at representative temperatures to which this scaling can be applied)

A_{d1} = area of the detector for which the dark current is to be estimated

A_{d2} = area of the detector from which the scaling is derived

K = another constant within the temperature range

T = temperature in Kelvins

Reference

1. Rule partially developed by Dr. Robert Martin and Dr. George Spencer, 1995.

Additional Sources

J. Miller. 1994. *Principles of Infrared Technology.* New York: Van Nostrand Reinhold, 137–138.
P. Norton. 1991. Infrared image sensors. *Optical Engineering* 30(11), 1649–1662.

NOISE BANDWIDTH OF DETECTORS

Subject: Detectors, physics, engineering, systems

The Rule

When unknown, the noise bandwidth of a photo-detector is commonly assumed to equal one divided by two times the integration time or:

$$N_b = \frac{1}{2t_i}$$

where N_b = noise bandwidth
t_i = integration or dwell time

Basis for the Rule

This rule is based on simple electrical engineering of photo-detectors and simplification of circuit design.

Cautions and Useful Range of the Rule

This rule is highly dependent on detector material, architecture, and optimal filtering of noise. It assumes a rectangular pulse and a value for the noise cutoff or 3 dB down from the peak.

The rule does not always properly include all readout and preprocessing signal conditioning effects, which can vary from $1/t_i$ to $1/4t_i$.

Usefulness of the Rule

The rule is useful for estimating a detector's noise bandwidth for D* or NEP calculations.

Notes and Explanation

The noise bandwidth of a detector is a function of signal processing, semiconductor physics, and readout architecture. A detector scanning a target will have a sharp increase in its output, which can be assumed to be almost a square wave in time. The response of the electronics (amplifiers and filters) to this will have a more gentle rise related to the inverse of the bandwidth. Similarly, a target entirely located on a pixel of a staring array provides a square pulse increase for the readout amplifier. For a well designed system, the electronics can be matched to provide a minimum noise bandwidth as approximated above.

PEAK VS. CUTOFF

Subject: Detectors, FPAs, systems

The Rule

A photovoltaic or photoconductive semiconductor detector's peak sensitivity is highest at a wavelength of about 10 percent less than its cutoff.

Basis for the Rule

This rule is founded on empirical observations, the state of the art, and simplifications of solid-state physics

Cautions and Useful Range of the Rule

The state of the art changes, so the situation can change.

The above rule is simply not valid for pyroelectric, ferroelectrics, quantum wells, Schottky barriers, or bolometers.

Usefulness of the Rule

This is useful for understanding, specifying, and estimating the cutoff wavelength at which a detector material should be made (specified).

Notes and Explanation

Usually, a classic semiconductor material has the characteristic that the maximum sensitivity is just lower than its 50 percent cutoff wavelength. Therefore, one should specify a cutoff of about 5 to 10 percent beyond the longest wavelength of interest. Frequently, vendors specify a "cutoff" as the 50 percent point but give the D* (or other sensitivity figure of merit) at the peak (see Fig. 6.1).

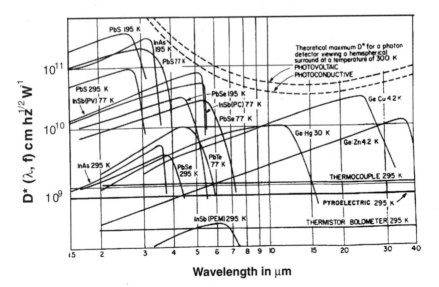

Figure 6.1 Wavelength vs. D* (courtesy of EDO/Barnes)

FPA COST

Subject: Detectors, FPAs, project management

The Rule

A focal plane's cost is dramatically increased by increasing the number of pixels, sensitivity, uniformity, or physical size. Of course, decreasing any of these rarely has an inverse impact on FPA cost.

Basis for the Rule

The above assertions come from empirical observations and the sorry state of the art.

Cautions and Useful Range of the Rule

This is a gross approximation by cynics; sometimes the cost can be reduced.

Usefulness of the Rule

The above generalizations are valuable for estimating cost impacts and underscoring the truth of FPA costs.

Notes and Explanation

Generally, the only way to reduce the FPA cost is to design your system around a standard FPA product.

DC PEDESTAL

Subject: Detectors, radiometry, systems, the sad facts of life

The Rule

Martin[1] points out that the signal one wants to observe is usually small compared to the electronic DC background. Therefore, detection of the signal requires sensing a small change in a large value.

Basis for the Rule

This rule is based on empirical observations, the fact that noise electrons often dominate the total number of electrons, and the state of the art.

Cautions and Useful Range of the Rule

This applies more to IR than UV or visible detectors.

Usefulness of the Rule

It underscores a serious hindrance in signal detection, and the length to which electronic engineers must go to enhance the signal.

Notes and Explanation

Martin points out that detection of a desired signal is often analogous to observing grass growing on the top of the empire state building (see Fig. 6.2). Typically, the signal (grass height) coming out of the detector is a minute voltage or current that frequently rides on a large level (the Empire State Building). The ratio between the two can easily be a 100:1, and sometimes 10,000:1. This large "pedestal" is (statistically) relatively constant and can be subtracted by a host of methods. However, this large "bias" eats up dynamic range and well capacity, makes readout and analog circuitry difficult to design and implement, and adds noise.

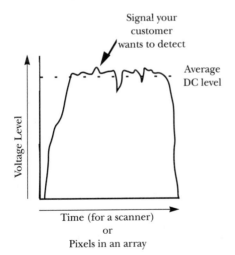
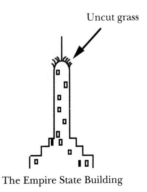

Figure 6.2 Signal Detection Analogy

This large level, relative to the signal, results from a myriad of sources including voltage sources, preamps, noisy resistors, background shot noise, dark current, 1/f effects, Johnson noise, and clutter. Scientific instruments frequently alternate between a view of the scene and a that of a known reference source ("chop") to reduce these effects. FLIRs often view a known reference and "characterize" the detectors several times per second, and signal processors can perform an AC coupling to effectively ignore this large "DC" level.

Reference

Rule provided by Dr. Robert Martin, 1995.

DEFINING BACKGROUND-LIMITED PERFORMANCE FOR DETECTORS

Subject: Detectors, system sensitivity, systems, radiometry

The Rule

The approximate background at which above, a detector can be considered background limited in performance (BLIP) is:

$$E_B = \frac{2.5 \times 10^{18} \lambda \eta}{D^{*2}}$$

where E_B = background radiation in watts
λ = wavelength in microns
D^* = specific detectivity in cm $Hz^{1/2}/W$ (Jones)
η = quantum efficiency

This can be rewritten for a photon flux above which the detector is considered BLIP:

$$\phi_B = 1.3 \times 10^{37} \left(\frac{\lambda}{D}\right)^2 \eta$$

where ϕ_B = background flux in photons/cm^2/s

Basis for the Rule

This rule is based on basic radiometry.

The constant of 2.5×10^{18} is the reciprocal of two times the Planck constant multiplied by the speed of light and adjusted for units of microns for the wavelength. The equations differ by the units in which the background is given.

Cautions and Useful Range of the Rule

One must make sure that D* and quantum efficiency are for the same wavelength of interest. This is based on a detector with a 2π steradian field of view (no cold shielding). Cold shielding (and cold filtering) will reduce background flux and improve the required performance to reach a BLIP condition. (See the "Sine Rule of Cold Shields" in Chapter 5.)

Usefulness of the Rule

It is useful in estimating the irradiance or flux for which a detector becomes BLIP.

Notes and Explanation

BLIP may not be achieved if the background is low, even for reasonably high R_oA products. Once the D* has been evaluated, based on a R_oA for a given temperature, a background may be estimated for which the background noise exceeds that of the detector, and the detector becomes BLIP. Once D* has been estimated, one of these equations for the BLIP level background may be employed.

SPECIFYING 1/f NOISE

Subject: Detector, systems, performance, specifications

The Rule

Spencer indicates that frequency where the 1/f noise equals the white noise (e.g., caused by shot noise, dark current, and so on) can be closely approximated by

$$f_o \leq \frac{(\text{fr})/2}{\ln\left(T_o[\text{fr}/2]\right)}$$

where f_o = "break" frequency (or "knee") of the 1/f noise, that is the place where it crosses the white noise power spectral density (PSD), in hertz (see Fig. 6.3)

  fr = frame rate in hertz

  T_o = observation time in seconds

T_o is the time in which you can allow the 1/f noise to grow and add to fixed pattern noise. For most systems, this will be the time between updates of the processing normalizing coefficients with a blackbody reference source. For many commercial cameras, this is the operation time (i.e., from turn-on to turn-off).

Basis for the Rule

This rule is based on simple noise analysis, and setting the 1/f noise equal to the white noise.

By definition, these types of noise are the standard deviation of the temporal fluctuations in the total signal. Often, the total signal level is relatively constant, and can be considered a *DC pedestal* (see "DC Pedestal Rule"). The average of this white

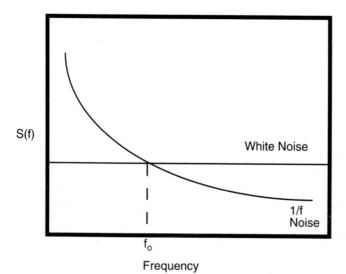

Figure 6.3 Noise Power Spectral Density

noise can be subtracted away. What is left are the variations (caused by targets, clutter and noise), which are aptly expressed either as a standard deviation and/or variance. Spencer[1] indicates that the variance (square of the standard deviation) of the white noise can be expressed as:

$$\sigma_w^2 = \int_0^{f_h} A \, df = (A) \times \frac{1}{2t_i} = \frac{A \text{ fr}}{2}$$

where σ_w = component of the white noise
$\qquad f_h$ = high-frequency cutoff, which is the noise bandwidth, $\approx 1/2t_i$
$\qquad f$ = frequency of interest
$\qquad t_i$ = integration time
$\qquad A$ = a constant
$\qquad$ fr = frame rate

Many forms of semiconductor electronics (especially detectors) tend to experience a slowly varying noise source that constantly increases with amplitude, called $1/f$ noise. This $1/f$ noise can be approximately (but still very accurately) expressed as:

$$\sigma_f^2 = \int_{f_l}^{f_h} \frac{df}{f/f_o}$$

$$\sigma_f^2 = A f_o \ln\left[\frac{T_o \text{ fr}}{2}\right]$$

where σ_f = component of the $1/f$ noise
$\qquad f_l$ = the low-frequency cut-on, which is observation time or the time between renormalizing the system (whichever is less)
$\qquad A$ = a constant

Spencer indicates that the limits on the integral are somewhat controversial. But, generally, for $1/f$ noise, these are set to the high-frequency cutoff and the low-frequency cut-on.

Setting these noises to equal levels and doing some math will result in the above rule.

Cautions and Useful Range of the Rule

The actual $1/f$ noise depends on detector material and the architecture of the electronics and assumes that the white noise bandwidth is equal to $(1/2\, t_i)$.

This rule sets the $1/f$ noise equal to the white noise. In some cases, it may need to be set even lower. For a system where $1/f$ noise tends to dominate, one should set it to less than the white noise (e.g., add a "3" to the denominator of the above equation to make it about 11 percent of the white noise).

Usefulness of the Rule

The above rule is quite useful for determining the specification for the $1/f$ noise frequency and underscoring the importance of properly defining $1/f$ noise.

It is also useful when attempting to calculate the effects (or importance) of $1/f$ noise from a given detector.

Notes and Explanation

Many detector materials (PbS, PbSe, HgCdTe, InSb) exhibit an unpleasant and disturbing noise that increases with time; this traditionally has been called *1/f noise*. Electrical engineers frequently deal with this in circuits. Its origins are still somewhat mysterious, but for detectors it seems to be a consequence of surface effects and is sensitive to the kind of passivation used.

Often, system designers specify the 1/f noise to have a "knee" at or below the frame rate. However, this may lead to unexplained noise after the hardware is built. If the 1/f knee is at f_o as defined above, then the RMS of 1/f noise just equals the RMS of the white noise. This results in a specification for the break frequency less than the frame rate. If you do not want to impact system noise, then the noise from 1/f effects should be $\approx 1/2$ or less than the of the noise from other sources, which may drive it even lower.

The deleterious system performance effects of 1/f noise can be mitigated by AC coupling, special signal processing, renormalizing the detector by forcing it to view a known blackbody source, or reducing the integration time.

Example

Assume a 30-Hz video system is updated with a radiometric reference only once per minute. From the above rule, the 1/f noise knee frequency characteristic of the detector should be specified as:

$$f_o \leq \frac{30/2}{\ln\left[60\left(30/2\right)\right]} = \frac{15}{6.8} = 2.2 \text{ Hz}$$

Again, note that 2.2 Hz is much lower than what might be expected, as the 30 Hz frame rate is sometimes assumed to be the break frequency. It should be noted that the "ln" in the denominator causes this to change only slightly with different observation times. For instance, if one only wished to update the above example every hour, the break frequency would only decrease to only 1.4 Hz.

Reference

1. Rule provided by Dr. George Spencer, 1995.

TDI SENSITIVITY IMPROVEMENT

Subject: Detectors, systems, radiometry, FPAs

The Rule

When finally implemented in a system, the gain from "n" elements in TDI is really more likely increase to the one-third root of the number of elements, rather than the square root.

Basis for the Rule

This rule is based on generalization of empirical observations coupled with a snide remark.

Cautions and Useful Range of the Rule

This assume no Herculean efforts in the manufacture of the FPA, servo system, or optics. Sometimes a system may approach the theoretical gain, but this is rare.

Usefulness of the Rule

Rapid estimates of the real (pessimistic) expected gain of adding TDI to a system underscore that adding more TDI elements is not always a panacea.

Notes and Explanation

Classic theory indicates that the sensitivity gain from adding TDI should vary as the square root of the number of TDI channels. Some FPA manufacturers, and a few systems, have almost achieved a "root-n" improvement. However, in practice, the gain is sometimes much less.

To realize the theoretical gain, the target must be precisely moved across the TDI elements by the sensor with zero error in pointing and scanning, FPA timing, optics distortion, optics roll, target signal variation, and FPA uniformity. The more TDI elements that exist, the harder it is to minimize these errors. In a real system, none of these errors is ever zero. For actual hardware, the gain from TDI is usually somewhere between the square root of the number of elements and the one-third root in the number of elements.

There was also a "hidden agenda" behind the development of HgCdTe TDI arrays, which does have a favorable system impact. The TDI process requires that the linear array be composed of rows of several elements in width (the in-scan direction). For a four-deep TDI, four in-scan elements are required. Many manufacturers took the opportunity to place more than the required "n" elements and then allow the user to employ only the "n" best. There are some 480×4 arrays that actually have 6 (or even 8) elements in the in-scan direction, allowing the user to select the best 4 of the row. This makes manufacturing easier, diminishes pixel outages, and provides a brute-force uniformity correction.

TEMPERATURE LIMITS ON DETECTOR/DEWAR

Subject: Detectors, systems, environments

The Rule

Generally, dewars containing photovoltaic HgCdTe FPAs should never experience temperatures in excess of 90° C. Dewars for other FPA materials should never exceed 100° C.

Basis for the Rule

This rule is derived from the state of the art in FPA/dewar production and generalization of normal bake-out temperatures and procedures.

Cautions and Useful Range of the Rule

The reader should be aware that higher exterior temperatures are tolerable for very brief periods, as these will not have time to heat up the critical areas.

This does not consider FPAs and dewars that are specially made to accommodate higher temperatures. On the other hand, maximum safe temperatures can be substantially lower if bake-out temperatures were lower.

Photoconductive materials and single-element detectors are often more tolerant of high temperatures.

Usefulness of the Rule

It is useful to readily identify questionable environmental specifications. It is also useful to limit the environments to ensure that a system will work and to illustrate the fragile nature of FPAs and dewars.

Notes and Explanation

Focal planes and detectors tend to be fragile devices that rarely can survive high temperatures without degradation. Likewise, the optics and coating inside a dewar can rarely survive temperatures in excess of the above limits. Moreover, cleanliness is the key to maintaining dewar vacuum integrity and, hence, life. Dewar manufacturers "bake out" every component to high temperatures, and the entire dewar assembly to 70 to 110° C. If the dewar is ever heated to a temperature near its bake-out temperature, contaminants may be released that will limit performance and lifetime.

WELL CAPACITY

Subject: Detectors, FPAs, noise in systems

The Rule

The well capacity of a readout device can be assumed to have a maximum value of about 25,000 electrons × the area of the pixel in square microns.

Basis for the Rule

This is based on the state of the art and on available 40- and 50μm square pixels with currently available deep wells. It assumes TTL bias (or less).

Cautions and Useful Range of the Rule

This assumes deep wells. Many devices (especially visible CCDs) hold less, with standard silicon readouts, which one can only assume has a capacitance of 1.6 pF at 5 V.

It does not account for special charge-skimming electronics that can increase the "effective" size of a well by subtracting some of the charge build-up and, hence, increase the allowable integration time for systems limited by well size.

Any pixel has some associated overhead electronics that take real estate away from the well capacitors. Usually, they occupy a 5×5 to a 10×10 micron area. Typically, these lines, feeds, and control circuits do not change with pixel size. Therefore, as pixels get smaller, the above rule overpredicts the well capacity.

Caveat emptor.

Usefulness of the Rule

It can be used for calculating well capacity in any spectrum, given a silicon readout multiplexer.

Notes and Explanation

Under normal bias conditions, readout structures (e.g., CCDs) can contain in a capacitor about 25 to 30 thousand electrons per micron squared. Each pixel can accommodate an area equal to less than the pixel for charge storage. In general, a 50-micron readout unit cell should be able to hold around 50 million electrons from the detector pixel.

Quantum Efficiency for Schottky Barrier Detectors, or the Fowler Equation

Subject: Detectors, Schottky barrier detector response, system sensitivity

The Rule

Knowing the detector's Fowler emission constant (C_1) and cutoff permits the calculation of the quantum efficiency for any Schottky Barrier at any wavelength as follows:

$$\eta(\lambda) = \frac{1.24\,C_1}{\lambda}\left(1 - \frac{\lambda}{\lambda_c}\right)^2$$

where $\eta(\lambda)$ = quantum efficiency at a given wavelength
C_1 = Fowler emission constant (usually between 0.2 and 0.4 for Pt:Si)
λ = wavelength in microns
λ_c = device cutoff wavelength, in microns, which is determined by the Schottky barrier potential

Basis for the Rule

This is based on semiconductor physics as described below.

Cautions and Useful Range of the Rule

One must be sure to use the correct C_1 and cutoff wavelength.

This does not account for specially tuned pixel cavities, microlenses, and other tricks to slightly increase the effective quantum efficiency.

This rule applies to Schottky barrier detectors only (e.g., Pt:Si, Ir:Si, Pd:Si) and simply does not apply to others (e.g., InSb, HgCdTe, QWIPs, and so on).

Usefulness of the Rule

This is quite useful for determining the quantum efficiency of a Schottky barrier detector at a given wavelength or across a given bandpass.

Notes and Explanation

Schottky barrier detectors experience the characteristic of decreasing sensitivity with increasing wavelength. A Schottky barrier's total sensitivity should be defined by parameters that characterize the internal photoemission (via the Fowler equation) and residual noise by the detector and multiplexer. These are C_1, cutoff wavelength, and total noise. The above Fowler equation determines the quantum efficiency based on cutoff and C_1. For the internal noise, use the manufacturer's data.

Sources

A. Fowler et al. 1990. A 256 by 256 hybrid array for astronomy applications. *Proc. SPIE*, vol. 1341, 52–55.

F. Shepard. 1985. Silicide infrared sensors. *Proc. SPIE*, vol. 930, 2–7.

W. Kosonocky et al. 160 by 244 element PtSi Schottky-barrier IR CCD image sensor. *IEEE Transactions on Electron Devices* 32(8).

J. Miller. 1994. *Principles of Infrared Technology*. New York: Van Nostrand Reinhold, 133–134.

CHARGE TRANSFER EFFICIENCY

Subject: Detectors, charge coupled devices

The Rule

Charge transfer efficiency usually increases as the temperature is lowered, and decreases as accumulated total dose (in rads) increases.

Basis for the Rule

The above assertion is from empirical observations, solid-state physics, and the state of the art of the mid-1990s.

Cautions and Useful Range of the Rule

The effects are nonlinear, so care should be used.

Usefulness of the Rule

This is useful for understanding the effects on charge coupled devices (CCDs) and in degrading specifications for temperature and nuclear effects.

Notes and Explanation

A CCD operates by transferring the charge in one pixel across a row through the other pixels. A large CCD may transfer some of the charge several thousand times before it reaches an amplifier. The efficiency of the transfer is critical to the prevention of shadowing effects and reduced signal to noise. As a CCD is cooled, its transfer efficiency usually increases, and its inherent noise decreases; this is why some very large and highly sensitive CCDs operate at reduced temperatures.

As CCDs are exposed to nuclear radiation, their performance decreases. There is a total-dose deleterious effect. Insulating oxides break down, so shorting occurs. Additionally, the wells tend to fill up with noise-generated electrons, and the charge transfer efficiency is reduced. CMOS structures do not have this problem but suffer from other negative effects when exposed to radiation.

NONUNIFORMITY EFFECTS ON SNR

Subject: Detectors, E-O system design, sensitivity analysis

The Rule

The maximum useful signal-to-noise ratio (SNR) is closely related to the reciprocal of the nonuniformity in the pixels in an FPA.

$$SNR_{max} \propto \frac{1}{U}$$

where SNR_{max} = maximum attainable SNR

U = residual (after processing) nonuniformity (or tolerance of nonuniformity) in decimal notation (e.g., 3 percent = 0.03)

Basis for the Rule

This rule is based on analysis of the uniformity effects on staring array systems and typical image processing techniques. Generally, staring arrays produce a fixed pattern noise that is a result of pixel-to-pixel sensitivity and noise variations. The original equation in Money's paper[1] has $SNR_{max} = 1/U$.

Cautions and Useful Range of the Rule

One must use the residual nonuniformity after all focal plane corrections are applied. (Processing can reduce residual nonuniformity 10 to 100 times!)

The equation provides the maximum SNR, not the SNR that the camera actually will have.

The rule is valid for high background conditions with low uniformity for staring focal plane arrays and desired large SNRs.

Some algorithms and instruments that do not produce a display for humans are less affected by fixed pattern noise and nonuniformity.

The estimate of the nonuniformity depends on the difference between the scene temperature and the correction points (the flux levels of the electronic normalization). These should be as close as possible (see Fig. 6.4).

This assumes that the pixel can be corrected. Unfortunately, many materials exhibit pixels that defy correction, regardless of the number of points of the correction. Usually, these are considered "dead" or "out-of-specification" pixels and are frequently called *blinkers*. They have the irritating property of blinking on and off in the scene and must be accommodated for in the image processing for a useful scene. To make matters worse, sometimes the "blinking" pixels change each time the array is turned on. These often exhibit excessive noise and, as Schulz and Caldwell[2] point out, the noise is often of the $1/f$ type.

Usefulness of the Rule

This rule may be used to underscore the importance of uniformity and uniformity correction and to estimate maximum attainable SNR based on uniformity (assuming that nonuniformity is the limiting factor in system performance).

Notes and Explanation

Early staring focal plane arrays in the infrared had great variation from one pixel to another. This variation (or nonuniformity) resulted in a strong noise source depen-

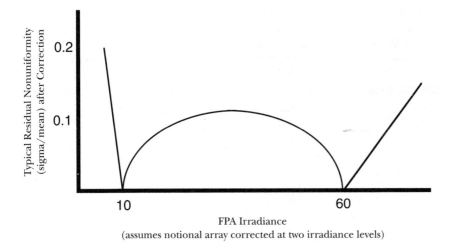

Figure 6.4 Nonuniformity as a Function of Flux for a Typical Focal Plane Array

dent on the background, called *fixed pattern noise*. In many camera situations, it is the dominating noise source. This has been used to justify the application of lower sensitivity (but higher uniformity) arrays (e.g., Pt:Si).

FPA manufacturers have been improving the uncorrected uniformity as well as the corrected uniformity. Often, they measure (and quote) the sigma (standard deviation of the gain, offset, or noise) divided by the mean for a given impinging flux or blackbody temperature. Then, after a correction at some given flux (or commonly stated as blackbody temperature through a given telescope and optical bandpass), the variance is greatly reduced. Reduction factors of 10 to 100 are common for two- or three-point correction. However, the corrections are at specific points in the impinging flux; in between these points, uniformity decreases, as can be seen by the notional graphic. The graphic indicates that a representative FPA's uniformity can be close to perfect at any given flux but degrades as the scene moves away from the points for which the corrections were made. The important point is to correct with a reference flux as close as possible to the expected scene flux, and to correct often. If the scene is spatially or temporally varying, then correct at as many flux levels as possible (four fluxes are better than three, which are better than two).

Example

One may have the necessary NEDT or sensitivity to detect a phenomena with a given signal to noise (say, 200). However, if fixed pattern noise (usually caused by nonuniformities in the FPA) is not considered, the results may be disappointing. Having a raw (uncorrected) variation from pixel to pixel of 10 percent, we would find the signal to noise reduced to 1/0.1 or 10. With the modest electronics and common normalizing procedures, the final corrected nonuniformity usually can be reduced to less than 1 percent. Therefore, one may notice an irritating fixed pattern noise that can limit the signal to noise to (1/0.01) or 100. Nevertheless, this is less than the ratio of 200 that one might first assume.

References

1. J. Mooney et al. Responsivity nonuniformity limited performance of staring infrared cameras. *Optical Engineering*, vol. 28, 1151–1161.
2. M. Schulz and L. Caldwell. Nonuniformity correction and correctability of infrared focal plane arrays. *Infrared Physics And Technology* vol. 36., 763–777.

Additional Source

Rule partially developed by Phil Ely. 1995.

HgCdTe x CONCENTRATION

Subject: Detectors, systems, manufacturability, cost estimates, HgCdTe wavelength cutoff determination

The Rule

The energy bandgap (and hence the wavelength cutoff) of a mercury cadmium telluride detector can be estimated from the operating temperature and the x concentration by:[1]

$$E_g = -0.295 + 1.87x - 0.28x^2 + [(6 - 14x + 3x^2)(E-4)]T + 0.35x^4$$

alternatively:[2]

$$E_g = -0.32 + 1.93x + 5.35 \times 10^{-4} T(1 - 2x) - 0.81x^2 + 0.832x^3$$

where E_g = energy bandgap in electron volts
x = material's x; that is, the relative concentration between Hg and Cd in the alloy, where x is the decimal concentration of Cd
T = temperature in Kelvins

Remember that the cutoff wavelength is related to bandgap by $\lambda = 1.24/E_g$.

Basis for the Rule

This rule is based on empirical observations and semiconductor physics. The amount of mercury with respect to cadmium determines the energy gap and, therefore, the long wavelength cutoff.

Cautions and Useful Range of the Rule

The rule provides estimation only but seems to track most companies' HgCdTe material to within ±1 micron in cutoff. However, the author has also found at least one device that has a cutoff of 2.5 μm lower than predicted from the above equations.

The reader is cautioned that this is the "detector material" cutoff, not someone's spectral filter cutoff. Also, this is detector cutoff, not peak wavelength, so the FPAs should be operated at 0.1 to 0.2 (or more) microns less than this.

The rule is valid from x concentrations of ≈0.15 to 0.45, for temperatures of, say, 50 to 200 K and cutoffs from about 2.5 to 15 μm.

There are many different equations to estimate the cutoff as a functions of x. Almost every detector manufacturer or systems house has its own. Generally, they all take the same form as the above equations, with some different coefficients.

Usefulness of the Rule

The rule is useful for estimations of the cutoff at a given mixture and temperature, demonstration of the effects of temperature and x on an array's properties, demonstration of the required degree of control for x based on a given uniformity, and estimating how the change of temperature of an existing array affects spectral response.

Notes and Explanation

Detectors made from mercury, cadmium, and telluride (HgCdTe) have their energy bandgap dependent on the concentration of Cd to Hg (called x) and their operating temperature. Note that this rule predicts the energy bandgap.

The equations indicate that, as x is decreased, the wavelength that the detector will respond to is increased. Additionally, as a slight change in x occurs, the cutoff wavelength will vary much more as a function of the wavelength, with the longer wavelengths being much more sensitive to this change in x. This is why process control is so critical (and difficult) for long-wavelength devices but relatively easy for short-wave and mid-wave devices.

References

1. J. Chu, Z. Mi, and D. Tang. 1991. Intrinsic absorption spectroscopy and related physical quantities of narrow-gap semiconductors $Hg_{1-x}Cd_xTe$. *Infrared Physics, vol.* 32, 195–211.
2. G. Hansen, J. Schmidt, and T. Cassleman. 1982. Energy gap vs. alloy composition and temperature in $Hg1-xCdxTe$. *Journal of Applied Physics, vol.* 53, 7099–7107.

Additional Source

J. Miller. 1994. *Principles of Infrared Technology.* New York: Van Nostrand Reinhold, 135–137.

7

Fiber Optics

Revolutions and fortunes in photonics are rare, but one could say that optical fibers have certainly caused more than their share. Their enormous transmission capabilities and unique optical properties (e.g., allowing light to bend) have changed many parts of the electro-optics environment. As commonplace as fiber optics appear to be today, as recently as the 1970s, they were considered exotic. In fact, it was an adventure when several firms first offered scientific fiber optic test apparatus in the late 1970s.

An optical fiber works by exploiting the total internal reflection that results at the interface between two materials of different indexes of refraction. An optical fiber is constructed so that the fine strand of transparent material is cylindrically "clad" with a lower-index material. That is, the core has a higher index of refraction than the cladding such that it provides total internal reflection. Usually, an additional surrounding coating is provided for protection and strength. Typically, this is referred to as a "buffer" or "protectant." In practice, fibers can be grouped into coherent "bundles" that can be made to transmit an image, or they can be grouped into incoherent bundles that simply transmit light energy.

Fibers are frequently classified by the number of modes they transmit (see Fig. 7.1). A *single-mode* fiber is efficient only for a single mode and is frequently used in communications that require long distances and high data rates. A graded index *multimode* fiber transmits several modes and is constructed with a radially varying index of refraction. There are two kinds of these fibers: a *step index* fiber, which has a flat index profile for the core, and a regular *graded index* fiber, which has a increasing index for the fiber. Multimode fibers allow the transmission of many modes. However, a pulse is lengthened as it travels down the fiber due to the different time delays of the various modes. Such broadening limits the pulse rate and, hence, communication rate. This type of fiber is the easiest to successfully inject (in fiber jargon this is called *launch*) a pulse into, but it suffers from lower data rates.

In 1854, John Tyndall demonstrated to the Royal Society that fiber optics hold promise, and in 1880 none other than Alexander Graham Bell proposed telecom-

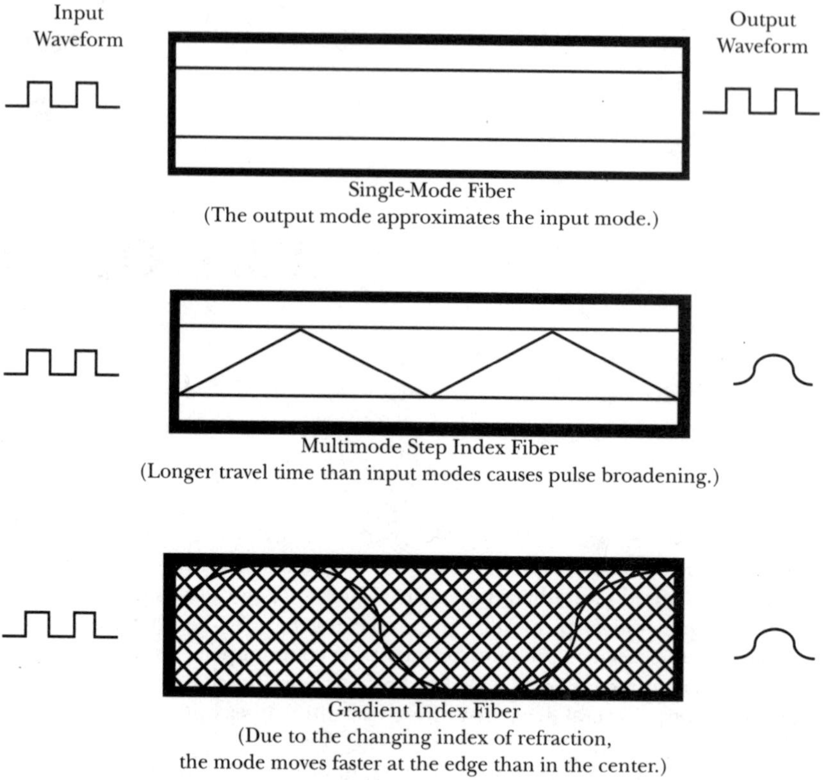

Input Waveform

Output Waveform

Single-Mode Fiber
(The output mode approximates the input mode.)

Multimode Step Index Fiber
(Longer travel time than input modes causes pulse broadening.)

Gradient Index Fiber
(Due to the changing index of refraction,
the mode moves faster at the edge than in the center.)

Figure 7.1 Fiber Optic Classification by Transmission Mode

munications via light waves. Since then, incremental improvements have transformed fibers from the realm of the physics laboratory to a household necessity.

Biomedical uses are also key to fibers' continuing advancements. Fibers are frequently used for visual exploration of the gastrointestinal tract and bladders, and for laparoscopies. For example, many years ago, knee surgery was expensive, and recovery periods were long. Now, with optical fibers, knee surgery is often an outpatient procedure, with full recovery within a few weeks. In the near future, fiber optics will be employed to examine arteries and sinuses (get ready for them going up your nose!) and are likely to be used for eye repairs.

Optical fibers are frequently employed in situations where electromagnetic induction (or crosstalk) is a concern, such as satellites, aircraft, and airports. New military aircraft frequently employ fiber optic data communications, and commercial airplanes will be retrofitted with optical fiber networks for entertainment and data communication for the passengers.

Optical fibers are also being used as a analytical tools for diagnosing stresses and faults in materials. They can be used to create "distributed aperture" systems (that is, employing a single sensor with multiple apertures that usually view different fields) and aimpoint fuses.

Optical fibers have not been utilized extensively in the ultraviolet or infrared parts of the spectrum due to their high absorption and relatively high material cost. However, recent advancements in fluoride glasses, chalcogenide materials, and hollow waveguides have opened this technology for (at least) the mid-infrared. Some time in the future, large-scale production costs in the infrared are likely to be about $1/m, whereas it is currently less than $0.10/m for the visible, and several tens of dollars per meter in the IR. Although currently uncommon, UV and IR fibers will eventually allow for distributed aperture systems, complex scene generation, and other advancements.

Finally, fiber optics are currently revolutionizing the transmission of information. The 1970s saw the development of low-cost diode lasers, and in the 1980s, manufacturing techniques allowed the low-cost mass production of fibers. Optical fibers can be used with lasers or light emitting diodes (LEDs) to carry enormous amounts of information as compared to conventional electrical systems using copper wire. Fiber optics represent a key enabling technology for the "information superhighway." Each of our lives will be transformed by the low-loss, high-bandwidth transmission of information via fibers. Hold on tight—here comes the future (on optical fibers).

For the reader who is interested in more details, several books can be found in the bibliography of this book (such as Kapany's *Fiber Optics*). If the reader is not familiar with fiber optic technology, the authors suggest first reading the material in these classic optics books to gain a familiarity with the basics. Excellent introductory discussions can be found in Hecht's *Optics* and Driscoll's *Handbook of Optics.*

SIZING A FIBER

Subject: Fiber optics, optics

The Rule

Usually, the core diameter of a multimode fiber ranges from 2 to 1000 μm, and most are commonly found in the 10 to 200 μm range. Also, a fiber's outer diameter is typically 1 to 10 times the core diameter.

Basis for the Rule

This rule is based on currently produced fibers. The state of the art changes, and there is some desire for smaller fiber diameters.

Cautions and Useful Range of the Rule

Obviously, this depends on the state of the art in the early 1990s. Also, custom-made fibers can be of much different sizes and ratios, and at least one of the standard sizes does not fit this rule.

Usefulness of the Rule

The above is useful for estimating available fiber sizes and understanding what sizes to expect.

Notes and Explanation

Dimensions of step-index fibers (see Fig. 7.1) vary greatly, depending on the intended application. Various fiber manufactures have developed standard sizes. The above rule is based on currently available fibers in the early 1990s, and the small diameters do not necessarily follow this rule.

Standard sizes of core and cladding include:

- 8- to 9-μm core with a 125-μm cladding (and a 1300-nm wavelength single mode)
- 62.5/125 for the Fiber Data Distribution Interface (FDDI) standard
- 85/125 the typical European standard
- 100/140 typical US military standard

Sources

This rule was partially developed by Judy McFadden, 1995.

N. Lewis and M. Miller. 1993. Fiber optic systems. In *Electro-Optical Components,* vol. 6, ed. C. Fox, of *The Infrared and Electro-Optical Systems Handbook,* executive ed. J. Accetta and D. Shumaker, Ann Arbor, MI: ERIM, and Bellingham, WA: SPIE, 245.

Cladding with an index of refraction of $>n_1$

Core with an index of refraction of n_1

Protective coating

Figure 7.1 Step-Index Fiber

OPTICAL FIBER BANDPASS

Subject: Fiber optics, fiber systems

The Rule

A glass optical fiber will transmit wavelengths between 250 and 2000 nm (0.25 and 2 μm).

Basis for the Rule

Most fibers are made from glass, and nature decreed that glass will transmit only a limited range of wavelengths.

Cautions and Useful Range of the Rule

Not all fibers are made of glass. The rule does not apply to plastic fibers.

Usefulness of the Rule

If the light launched into the fiber is outside the range indicated, nothing will emerge from the fiber. Thus, this is useful to remember when doing system design.

Notes and Explanation

Usually, the glass used in optical fibers is quartz (silica) and must be ultra-pure for low-loss transmission. Wavelengths shorter than 250 nm are attenuated because of absorption and Rayleigh scattering, and those above 2 μm are attenuated due to infrared resonant absorption bands within the glass. Small concentrations of dopants (such as germanium, phosphorous, boron, or fluoride) are sometimes added to change the index of refraction.

It was in the 1960s when processes became advanced enough to routinely remove enough impurities to create a fiber with an attenuation of <20 dB/km. By contrast, the attenuation of common glass window material is much greater than 1000 dB/km.

Reference

Rule provided by Greg Ronan, 1995.

NUMERICAL APERTURE DETERMINES TOTAL INTERNAL REFLECTION

Subject: Fiber optics, optics

The Rule

$$NA = \sin A = (n_1^2 - n_2^2)^{1/2}$$

where NA = nominal numerical aperture
A = entrance (acceptance) angle
n_1 = index of refraction of the fiber medium
n_2 = index of refraction of the fiber cladding

Basis for the Rule

This rule is based on determining the numerical aperture (NA) of a fiber from Snell's law, basic trigonometry, and fiber optic theory.

Cautions and Useful Range of the Rule

Although it is frequently applied for rays across the fiber optic edge, it really is derived for the meridianal ray (or central ray) only.

Light must enter from a medium with an index of refraction close to 1 (e.g., air or space, but not an immersion oil).

The actual acceptance angle is usually not as sharply defined, as the rule contends, due to by diffraction, striae, and surface irregularities.

Usefulness of the Rule

This is useful in the design stages to estimate the acceptance angle of a fiber. Conversely, for a given fiber, this equation can indicate the acceptance angle for the launch.

Notes and Explanation

A ray that encounters the exterior fiber face, with an angle of less than or equal to the acceptance angle, will undergo total internal reflection when it encounters the difference in index of refractions between the cladding and the fiber media. The numerical aperture can be "tuned" for larger NA by making the difference between the core and cladding greater. But beware, because in a single-mode fiber, the NA will get larger as the core size becomes smaller.

DEPENDENCE OF NA ON FIBER LENGTH

Subject: Fiber optics, optical communication

The Rule

When the numerical aperture (NA) is measured for a long fiber, it is usually found to be less than a short segment of the identical fiber.

Basis for the Rule

This rule is based on waveguide physics and supported by empirical observations.

Cautions and Useful Range of the Rule

It depends on the state of the art in the early 1990s and is generally useful for comparing length differences of factors of two or more.

Usefulness of the Rule

This can be used to add margin to NA in the design process and is useful as a reminder that the NA will be less for long fibers.

Notes and Explanation

The above rule-of-thumb should be heeded when making measurements or determining specifications. This phenomenon results in the relatively greater attenuation of the high-order (high-angle) modes (or rays) within the fiber.

Sources

Technical Staff of CSELT. 1981. *Optical Fiber Communication.* New York: McGraw Hill, 267.

N. Lewis and M. Miller. 1993. Fiber optic systems. In *Active Electro-Optical Systems*, vol. 6, ed. C. Fox, of *The Infrared and Electro-Optical Systems Handbook*, executive ed. J. Accetta and D. Shumaker, Ann Arbor, MI: ERIM, and Bellingham, WA: SPIE, 245.

WAVELENGTH CUTOFF IN SINGLE-MODE FIBERS

Subject: Fiber optics, optical communication

The Rule

The following calculation defines the wavelength above which only the single (or lowest order axial) mode exists, thus providing the highest bandwidth for communications:

$$\lambda_c = \frac{2\,\pi\,a\,n_{core}\,(2\Delta)^{1/2}}{2.405}$$

where λ_c = cutoff wavelength for a single mode
   n_{core} = index of refraction of the core
     a = radius of the core

   $\Delta = \dfrac{n_{core} - n_{clad}}{n_{core}}$ where n_{clad} and n_{core} are defined as the indices of refraction

     of the cladding and core of the fiber

Basis for the Rule

This rule is based on diffraction theory supported by basic empirical observations.

Cautions and Useful Range of the Rule

This rule applies to single mode fibers only.

Notes and Explanation

The bandwidth of an optical fiber used for communications is often limited by the difference in the velocities of modes. The following calculation defines the wavelength above which only a single mode exists, thus providing the highest bandwidth. For wavelengths shorter than λ_c, more modes can exist and cause pulse stretching, thus reducing bandwidth.

Reference

N. Lewis and M. Miller. 1993. Fiber optic systems. In *Active Electro-Optical Systems*, vol. 6, ed. C. Fox, of *The Infrared and Electro-Optical Systems Handbook*, executive ed. J. Accetta and D. Shumaker, Ann Arbor, MI: ERIM, and Bellingham, WA: SPIE, 248.

CLEAVING FIBERS

Subject: Fiber optics, fiber cleaving, fiber systems

The Rule

To produce a perfectly flat cleave of an optical fiber, the crack speed must propagate more slowly than the speed of sound in the fiber.

Basis for the Rule

This rule is based on the physics of the circumstance. A cleave is initiated by lightly scratching the surface of the fiber. When the fiber is thereafter pulled or bent, a crack will originate at the scratch and propagate radially across the width of the fiber. The stress field within the fiber created by tension or bending determines the speed at which sound will propagate. If the crack exceeds this speed, the crack will suddenly change direction by almost 90°. This result is an excess of glass on one fiber (commonly called a *hackle*) and a shortage on the other fiber.

Cautions and Useful Range of the Rule

This rule should hold regardless of fiber material. Just be certain that you know the speed of sound in the fiber.

Usefulness of the Rule

Keep in mind that most bad cleaves are due to the initial scratch being too deep. Torsion will not change the speed of sound within the fiber, but it will produce non-perpendicular endfaces.

Notes and Explanation

It is easy to cleave an 80-μm diameter fiber, possible to cleave a 125-μm diameter fiber, and usually difficult to cleave >200-μm fibers. Again, torsion will produce a non-perpendicular endface. In fact, most commercially available angle cleavers rely on torsion. The endface angle is proportional to the amount of torsion.

Sources

Rule provided by Greg Ronan, 1995.
Wave Optics product catalog, 1995. Palo Alto, CA: Wave Optics, 23.

BENDING A FIBER

Subject: Fiber optics, optical communication, optomechanics

The Rule

Light fibers are flexible with curvature radii of 20 to 50 times their diameters when hot, during manufacture. After cooling, they are bendable to within about 300 diameters.

Additionally, Ronan[1] suggests that a glass optical fiber should not be coiled on a diameter less than 600 times the fiber diameter for long-term use, nor tighter than 200 times momentarily.

Basis for the Rule

This is based on empirical observations of the present state of the art of fiber optic materials.

Bending optical fibers can reduce their lifetime. During bending, the center of the fiber is stressed. Bending more tightly than these curvatures will damage the fiber. The inner edge of the fiber is under compression, while the outer edge is under tension. The compression and tension create micro-cracks on the surface of the glass, which weakens the fiber.

Cautions and Useful Range of the Rule

This rule depends on the maturity of the technology.

Various microbending equations can give a more accurate radius of curvature.

This rule is valid for optical fibers made of quartz (silica). Other materials have different mechanical properties.

Usefulness of the Rule

A quick check to verify the amount of curvature you can plan on bending a fiber optic, and understanding why you now have two pieces of fiber, when you previously had only one.

This rule is useful to consider anytime that your system requires bending a fiber. Bending more than the above will often result in over-stressing the glass and may cause a future break

Notes and Explanation

It is generally accepted that, when a fiber is bent, one must ensure that the radius of curvature of the bend is more than 500 (a safety margin) times the fiber diameter. Optical fibers are physical materials. When they are curved to an extent where mechanical stress dominates, they break. In general, with modern fibers, the above relation holds true.

In fact, an easy way to break a fiber into two pieces is to create a coil between your fingers and pull on one end. The fiber will not break until the coil diameter is small.

The attenuation of light in a fiber that has tight bends is greater for higher wavelengths (e.g., light of 1.55 µm has more loss than light from 0.85 µm for the same bend radius).

Example

An optical fiber is typically 125 μm in diameter. Therefore, the minimum coil diameter of the fiber should be 75 mm when the fiber is deployed or in storage. During handling, the fiber may be bent around much smaller diameters for a short time, with no tension applied.

Reference

1. This rule was partially developed by Greg Ronan. 1995.

BACKREFLECTION REDUCTION

Subject: Fiber optics, index matching, testing, systems

The Rule

To reduce backreflection (or generally increase attenuation), the fiber may be wrapped 3 to 5 times around a pencil. Alternatively, an index matching gel will reduce backreflection to below −60 dBm.

Basis for the Rule

Several "tight" turns of a fiber will result in the phenomenon of greatly increased attenuation. The sharp bending increases the angle between the traveling light and the cladding interface. If the bend is tight enough, it will be less than the critical angle, and light will leak into the cladding and be lost. A pencil provides a good radius of curvature for most typical fibers, with little probability of permanent damage.

The second part of this rule is based on matching the index of refraction of a gel to the fiber, thereby eliminating the glass-to-air interface.

Cautions and Useful Range of the Rule

The rule is based on current fibers and gels of standard sizes and typical pencils. If highly curved, fibers tend to break. When bending a fiber to a tight radius of curvature, do not apply tension, and heed the safety tips of the previous rule.

Usefulness of the Rule

This rule is useful for an understanding and estimation of the backreflections from fibers. It is also useful for reducing backreflections in testing and providing a quick, low-cost way of attenuating the light.

Notes and Explanation

"Backreflection is light reflected back toward the source relative to the forward signal, primarily caused by index of refraction discontinuities (Fresnel reflection). This effect can be controlled by connector endface preparation. In most systems, backreflection is not a significant problem. However, reflections back into a laser can result in output fluctuations and mode hopping. This is undesirable in some applications, especially high-speed and analog systems. Choosing the proper fiber polish will minimize system problems due to backreflection."[1]

Reference

1. Wave Optics catalog, *Single Mode Fiber Optic Products*, 2nd ed. 1994. Palo Alto, CA: Wave Optics, 4.

FIBER ADAPTER CLEANING

Subject: Fiber Optics, fiber adapters, systems

The Rule

If insertion loss suddenly increases with an adapter that has been used several times, the cause may be a dirty adapter.

Basis for the Rule

This rule is based current state of the art.

Cautions and Useful Range of the Rule

The cause is not always a dirty adapter, so beware. It may be a broken fiber.

Usefulness of the Rule

This rule is useful for troubleshooting an unexpected insertion loss.

Notes and Explanation

Fiber adapters are used to mechanically and optically connect two fiber optic connectors. Fiber optic adapters are almost always female to female while connectors are male.

Usually, a fiber adapter consists of two main elements: the alignment sleeve and the housing. The alignment sleeve is the critical part of the adapter, providing the alignment of the two connector ferrules. The alignment sleeve is usually a split "C" and is made from a hard, low-wear material. The housing provides the important mechanical connection that holds everything together.

The easiest way to clean adapters is with a pipe cleaner or alcohol-wetted cotton.

Source

1. Wave Optics catalog, *Single Mode Fiber Optic Products,* 2nd ed. 1994. Palo Alto, CA: Wave Optics, 7.

PULSE SPREADING PER KILOMETER

Subject: Fiber optics, optical communications

The Rule

According to Li,[1] a good figure of merit for using fibers in communications is the product of the bandwidth and the distance that it can carry a signal before pulse spreading leads to unacceptable error rate. This usually is between 2.0 and 2.5 GHz/km.

Basis for the Rule

This simple rule is based on empirical observations of current multimode fibers.

Cautions and Useful Range of the Rule

Because of the basis for the rule, it depends highly on the state of the art. It does not account for absorption losses in the fiber or couplers.

Usefulness of the Rule

This can be used to estimate the distance a fiber can carry a given data rate signal or data rate.

Notes and Explanation

The transmission bandwidth of a fiber is frequently driven by dispersion of the various modes. This arises because different modes have different group velocities. Dispersion also occurs in transmitting optical elements because of the variation of the index of refraction as a function of the wavelength of the light. Li also indicates that the modal dispersion can be minimized by a "near-parabolic refractive index profile in the fiber cross section." The greater the transmission length, the more pronounced this effect becomes. Current fibers support about 2.5 GHz transmission bandwidth for 1 km of fiber.

Additionally, for step index fibers, travel time variation between the axial mode and the higher-order modes of a cladded fiber cable can cause undesirable pulse stretching that may limit the data rate of the fiber. According to Hecht,[2] the time difference between the axial (shortest time) mode and the slowest mode is Δt, which can be expressed as

$$\Delta t = \frac{Ln_f}{c}\left(\frac{n_f}{n_c} - 1\right)$$

where Δt = time difference between modes
L = length of the fiber
n_f = index of refraction of the fiber
c = speed of light
n_c = index of the fiber cladding

Example

To transfer 500 Mb of information over 50 km, one can divide 2.5 GHz/km by 50 and estimate that the maximum data rate is 0.05 Gb (or 50 Mb) per second. Thus, the information can be sent over 50 km in only 10 s.

However, one should also check the Δt to verify the maximum data rate.

Pulses injected into the cable cannot exceed one pulse every $2\Delta t$ so that the pulses can be discerned at the receiver. A 50 km cable with a Δt of 10 ns can accept pulses at a rate of one per 20 ns, which is equivalent to 50 million pulses per second.

References

1. T. Li. 1985. Lightwave telecommunication. *Physics Today* (May).
2. Hecht, E. 1990. *Optics*. New York: Addison-Wesley, 175.

RESOLVING POWER OF A FIBER BUNDLE

Subject: Fiber optics, fiber bundles, optics

The Rule

The approximate resolving power (in lines per mm) of a fiber bundle is 500 divided by its diameter (in microns) or

$$R_{lpmm} = \frac{500}{d_f}$$

where R_{lpmm} = resolution in line pairs per millimeters
d_f = diameter of the individual fibers in microns

Basis for the Rule

This rule is based on empirical observations of fiber bundles based on the current state of the art.

Cautions and Useful Range of the Rule

This rule incorporates several assumptions, including high contrast, the use of common commercially available fibers (some high-tech ones will do better), adequate signal levels, static conditions (when scanning, it is more like $1200/d$), good fiber alignment, very closely packed fibers, and a resolution that is limited by the fibers.

Usefulness of the Rule

It is useful for estimating the proper diameter of a fiber for a given resolution required from a bundle and for determining the resolution of a fiber bundle.

Notes and Explanation

The resolving power of a fiber optic bundle is sensitive to many attributes as well as the conditions of the bundle and the measurement scenario. However, the above relationship provides a decent estimate of the expected resolving power.

A coherent fiber bundle can be defined as a bundle whose individual fibers preserve coherency (i.e., have the same spatial relationships to each other at both ends). When information (e.g., an image) is transmitted by such a bundle, it is transmitted in segments. Each segment is the size of the individual fiber diameters. Information (image content) smaller than a fiber diameter is not properly transmitted. Therefore, the resultant resolution is determined by the diameter of the individual fibers that compose the bundle. Moreover, the fiber bundle is synchronously scanned across the object and image planes, and the resolution of such a dynamic bundle is about twice that of a static bundle.

Sources

W. Seigmond. Fiber optics. 1978. In W. Driscoll and W. Vaughan, eds. *Handbook of Optics*. New York: McGraw-Hill, 13–10.

W. Smith. 1978. Optical elements, lenses and mirrors. In *The Infrared Handbook*, ed. W. Wolfe and G. Zissis. Ann Arbor: ERIM, 9–14.

MODE FIELD DIAMETER

Subject: Fiber optics, mode field diameter, fiber systems

The Rule

The mode field diameter (MFD) of a single-mode fiber is roughly six to nine times the operating wavelength.

Basis for the Rule

This rule is based on the physics of light propagation in a waveguide.

Cautions and Useful Range of the Rule

The MFD is dependent on several parameters, including numerical aperture (which decreases the MFD) and wavelength (longer wavelengths leads to larger MFDs).

Usefulness of the Rule

The MFD is needed to calculate the coupling efficiency from another fiber, a laser source, or a detector. It is also useful to consider it when sizing a fiber.

Notes and Explanation

It is generally desirable to have the MFD as large as possible so that maximum light can be coupled into the fiber. However, if MFD is too large, then higher-order modes can propagate. The physics of the light propagation determines the beam diameter.

The core of a fiber can be physically measured due to its higher index of refraction. In a multimode fiber, the light is confined to the core through total internal reflection. In a single-mode fiber, the light has a distribution that approximates a Gaussian (actually, a Bessel function in the core and a separate decaying exponential in the cladding). The mode field diameter is the point at which an equivalent Gaussian distribution has decayed to its $1/e^2$ points and is approximately 15 percent larger than the core diameter.

Sources

Rule provided by Greg Ronan, 1995.
Wave Optics product catalog. 1995. Palo Alto, CA: Wave Optics, 14.

Waveguide Rules

Subject: Fiber optics, optical communications, optoelectronics

The Rules

1. In a dielectric waveguide, all but the cylindrical symmetric modes (TE_{om} and TM_{om}) are hybrid modes.

2. The number of modes increases with the square of the fiber diameter.

3. Almost all modes mode have cutoffs that are a function of wavelength, fiber diameter, and fiber numerical aperture.

Basis for the Rules

These rules are based on simple physics, approximations to theory, and observations of the current state of the art.

Cautions and Useful Range of the Rules

These are valid for small fibers, where the transverse dimensions of the fibers approach the wavelength of the light. Therefore, this is especially of concern for infrared fibers.

Usefulness of the Rules

In optical communication fibers, the waveguide properties can become important, and single-mode fibers are sometimes required. These rules provide some simple relationships about the waveguide properties.

Notes and Explanation

The permissible modes within a fiber can be a critical design parameter and limit the transmission data rate. If the fiber's size is close to that of the wavelength of light being propagated, only specific distributions of the electromagnetic field will satisfy Maxwell's equations and thereby be allowable. This results in discrete waveguide modes. In large fibers, there are many modes that interact (usually to the detriment of the engineer) and overlap. Therefore, the above waveguide properties are often inaccurate for large fibers.

Sources

N. Kapany and J. Burke. 1961. *J. Opt. Soc. Am.*, vol. 51, 1067–1078.

E. Snitzer. J. 1961. *J. Opt. Soc. Am.*, vol. 51, 491–498.

W. Seigmond. Fiber optics. 1978. In W. Driscoll and W. Vaughan, eds. *Handbook of Optics.* New York: McGraw-Hill, 13-11.

8

The Human Eye

The function of the human eye has been a source of wonder for millennia. What is more, the advancements in modern science that have allowed us to fully characterize that which can be known about how the eye functions has led to many more questions than can be answered. By itself, the optical performance of the eye is quite poor. Even in persons with ideal vision, the image falling on the retina (the eye's detectors) has aberrations. Parallel lines appear to be curved toward and away from one another, the image is inverted relative to the objects in the scene (as in most other imaging systems) and in the middle of it all is a point at which there is no vision at all. Fortunately, the optical system is just the beginning of the vision system. Behind the eye is the most powerful computer known to man, and it corrects virtually all of these faults—although a pair of glasses is often invaluable in providing the finishing touch on the system.

Only because the eye is connected to the largest and most capable computer known to man—his brain—can we understand why vision of any quality occurs at all. The brain not only adapts to the aberrations of the normal eye, it can gradually accommodate insults that no EO instrument could deal with; application of inversion lenses cause the world to appear inverted for a time (usually on the order of days) but is eventually corrected by an inversion in the brain's interpretation of the image. Even highly distorting glasses are eventually overcome by processing that results from experience and practice. EO designers dream of a world in which the inevitable image distortions that occur in their systems could be removed by a computer system able to practice and learn. Perhaps a mature version of neural network technology will provide this capability.

The complexity of the functions of the eye should make it no surprise that many of the observed capabilities and limits of the eye can be described only empirically. Quantitative theories of eye function are limited to those that deal with the nature of parts of the system; rods, cones, nerve cells, and so on. The operation of the whole system is not well understood at all. Accordingly, many of the rules in this chapter are descriptive in nature; they do not, and cannot, include descriptions of why the phenomena occur.

As we find in this chapter, the eye is in nearly constant motion and yet, with the brain, it forms images of exquisite clarity. It is capable of diffraction-limited performance (although this is rare in the general population), in spite of all of the odd features of the system, such as the presence of a blind spot, aberrations, and defects that are present in the field of vision. It also has a huge field of regard, rivaling the performance of a wide-field sensor on a gimbal. It accommodates a wide range of illumination and reacts quickly to changing lighting conditions.

Perhaps the constant motion of the eye is necessary to eliminate the otherwise dark spot that would appear in the vicinity where the optic nerve appears in the retina (focal plane). Processing of the many images taken during eye motions, possibly relying on a method of image subtraction, results in a still picture. Indeed, in experiments, it has been shown that vision is suppressed when the eye is not allowed to move. Furthermore, experiments show that some structures in the optic nerve resemble nerve cells of the brain and may actually participate in image "preprocessing." This theory is supported somewhat by the fact that the time delay in communicating the presence of light into the brain is far longer than the observed reaction time to image motion. The possibility is that the optic nerve and the brain, working together, are able to keep up with the information flow.

The vision system made up of the eye and brain is covered in many texts at all levels. The interested reader can also find frequent contributions in this vital field in magazines such as *Scientific American*. Professional journals require a strong foundation in both optics and biology and are reserved for the expert. A particularly fascinating discussion of the function of the optical processing component of the brain is provided by Francis Crick in his recent book, *The Astonishing Hypothesis*. There is also an active research community in the military that is concerned with target detection phenomena and the way the mind/brain processes images. SPIE occasionally publishes compendia of these articles.

Optical Fields of View

Subject: Human eye, ergonomics

The Rule

The field of view of a *homo sapiens* approximates an ellipse 125° high and 150° wide. However, only a small portion of the field is in an area of acute vision. Moreover, in one minute, the eye can fixate on as many as 120 observations, which allows 0.2 to 0.3 s to fixate on each, with the remaining time used for eye movement.

Basis for the Rule

This is based on empirical observations of normal humans.

Cautions and Useful Range of the Rule

This generalization usually is correct, for healthy eyes. Generally, these numbers are a maximum.

Usefulness of the Rule

It is useful for quick determination of a human's approximate FOV. This is useful for designing displays, including head-up displays (HUDs), monoculars, virtual reality, and information systems applications.

Notes and Explanation

The above rule applies to normal humans under normal conditions. The portion of acute vision is determined by the fovea, which subtends a few degrees of the field. The density of cones is very high in the fovea. Rods dominate peripheral vision beyond a few degrees off axis. The outside portion of the field of view is used for orientation.

Source

B. Begunov, N. Zakaznov, S. Kiryushin, and V. Kuzichev. 1988. *Optical Instrumentation, Theory and Design.* Moscow: MIR Publishers, 73–74.

PUPIL SIZE

Subject: Human eye

The Rule

The size of the human pupil may be estimated from

$$D = 5 - 3 \tanh (0.4 \log B)$$

where D = pupil diameter in millimeters
B = luminance in candelas per square meter
tanh = the hyperbolic tangent function

Basis for the Rule

This rule fits a curve measured in human eye response to light (see Fig. 8.1).

Cautions and Useful Range of the Rule

This rule applies over the typical range of luminance to which the eye is most commonly exposed.

Usefulness of the Rule

The rule is useful for estimating a pupil's size under a variety of lighting conditions.
    This is also a consideration for display design; the geometry of the eye must be taken into account to avoid vignetting and to ensure full illumination as the eye moves.
    It is also useful in the study of ergonomic effects.

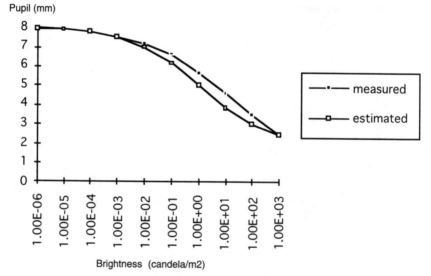

Figure 8.1 Brightness vs. Pupil Diameter

Notes and Explanation

This rule can be used in assessing eye safety issues, given that the size of the pupil determines the amount of energy received. Of course, the threat to the eye is that the light passing through the pupil may be significantly focused, resulting in very large energy densities at the retina (focal plane).

Sources

W. Driscoll and W. Vaughan, eds. 1978. *The Handbook of Optics*. New York: McGraw-Hill, 12-10–12-12.
The measured data in the graph is from F. Sears. 1958. *Optics*. Reading, MA: Addison-Wesley, 134.

Vision Creating a Field of View

Subject: Human eyes, human methodology for searching a field of regard

The Rule

Humans search a field of regard in a step-stare manner. An approximately 5° circular field is searched in about 3/10 of a second.

Basis for the Rule

This is based on empirical measurements of normal humans.

Cautions and Useful Range of the Rule

The rule is based on responses from people in good physical condition. Depending on the age distribution of the group, different fields of view and step times might apply.

Usefulness of the Rule

The rule allows for an estimate of the time needed to search a field of arbitrary size. The results are valuable in display design, symbol design, and other image processing tasks.

Notes and Explanation

Mathematically,

$$T = \frac{0.3 \, (FOR° x) \, (FOR° y)}{25}$$

where T = time in seconds to search a field
  $FOR° x$ = extent of the area to the searched (in degrees) in one direction
  $FOR° y$ = extent of the area to the searched (in degrees) in the other direction

Humans tend to step-stare across a scene to find an object. The time it takes is a function of the size of the field and the difficulty in finding the object. The above rule allows one to calculate the time it takes a person to search a field of view or display screen that does not contain complex or low-contrast images.

Reference

1. Burle Industries. 1974. *Burle Electro-Optics Handbook.* Lancaster, PA: Burle Industries, 121.

Visual Target Detection

Subject: Human eye, systems, ergonomics

The Rule

If a group of observers is given unlimited time to search for a target in a field, the fraction that find the target is called P_∞. The probability of detection for a group of observers varies with time t according to

$$P(t) = P_\infty [1 - \exp(-t/t_{fov})]$$

where t_{fov} = mean acquisition time if the target is found
t = time the group is given to find the target

Basis for the Rule

The rule is based on empirical observation.

Cautions and Useful Range of the Rule

P_∞ describes the fact that not all of those tested ever see the target, given an unlimited amount of time. Thus, $P(t)$ is the fraction of those tested, or who are working in the field, who are able to detect the target.

Usefulness of the Rule

This rule assists in determining the time that must be allowed for observing phenomena related to target detection. It is useful when designing experiments related to camouflage and other efforts to hide vehicles using paints and surface coatings.

Notes and Explanation

The value of t_{fov} is related to the probability that the target will be acquired in a single glimpse by

$$\tau_{fov} = \frac{t_f}{P_0}$$

where t_f = fixation time of the eye, which is described elsewhere in this chapter as being about 0.3 s
P_0 = the probability that acquisition occurs in one glimpse

Source

J. Lloyd. 1993. Fundamentals of electro-optical imaging systems analysis. In *Active Electro-Optical Systems*, vol. 4, ed. M. Dudzik, of *The Infrared and Electro-Optical Systems Handbook*, executive ed. J. Accetta and D. Shumaker, Ann Arbor, MI: ERIM, and Bellingham, WA: SPIE, 111.

SUPERPOSITION OF COLORS

Subject: Human eye

The Rule

A beam of red light overlaid with a beam of green light causes the eye/brain to sense yellow.

Basis for the Rule

The rule is based on human physiology.

Cautions and Useful Range of the Rule

The color sensed is dependent on the specific colors that are overlaid, but the statement above is the common experience of most people.

Usefulness of the Rule

This rule emphasizes the fact that human vision is complex and has some unexpected characteristics. Who could have guessed that the brain and eye, working together, would synthesize the same color that we get when we mix pigments of the two colors?

Notes and Explanation

This is one of many characteristics of the human eye that so far elude our explanation. For example, an even more fascinating extension of this rule has been pointed out by Crick.[1] He mentions that the color perception process works even if the colors mentioned in the rule are shown to the observer, one after the other, as short flashes. If the second color is not shown, then the observer sees only the first color. Therefore, we must conclude that the brain processes all of the information obtained in a period of time before concluding what has been seen. For example, one of the authors (Friedman) has had the opportunity to try to determine if the eye can sense color when exposed to extremely short light pulses. In the experiment, he used a nitrogen laser pumped dye cell to produce 5 ns pulses of a color not known to the subject. The color was actually green. Single pulses were then imposed on a non-fluorescent target, which was observed by the subject. Any subject could reliably tell the color of the pulse, even though conventional wisdom is that humans are limited in their ability to sense short light pulses to about 1/30 of a second (which is the underlying assumption in the creation of motion pictures).

Reference

1. F. Crick. 1994. *The Astonishing Hypothesis.* New York: Scribners, 72.

Additional Source

E. Hecht. 1990. *Optics.* Reading, MA: Addison-Wesley, 73.

MAXIMUM MAGNIFICATION CHANGES

Subject: Human eye

The Rule

The magnification change in an optical system with an output that is viewed by a human should not exceed a factor of 5 (and probably should be less than a factor of four) from one step to another as an optical system is stepped in or out.

Basis for the Rule

This rule is based on approximations of empirical observations of ergonomics.

Cautions and Useful Range of the Rule

This rule should be used only as a general guideline, as special circumstances or computer assistance could change this rule.

This rule really applies to a switchable discrete FOV, not a zoom; the continuous motion of a zoom usually allows a greater change in magnification without confusing the observer.

Usefulness of the Rule

It is useful in system design.

Notes and Explanation

It is useful to define the difference between *field of view* and *field of regard*. Field of view is the inherent angular field over which the eye can function without motion. Field of regard is defined as the total field of the eye, including the motions that can occur within the head.

An example of this effect comes from the world of forward-looking infrared systems. Most FLIRs have multiple fields of view that are user-switchable during operation. This gives a wide, general surveillance mode as well as a higher magnification and narrower field for targeting, designation, or detailed intelligence gathering. Often, a user activates a wide field of view to survey a large area and then switches to a narrow field to check "hot spots" and eliminate false alarms.

If the switch in FOV exceeds a magnification change of 4 or 5, the user often has trouble following the details of the image and has difficulty tracking the object of interest.

There is also a practical reason why optical steps larger than about 3 are uncommon. A large zoom capability means that some sort of line-of-sight stabilization must be built in or the field will have a lot of jitter that will prevent the user from viewing the scene successfully. On the other hand, a very wide field of view tends to have large optical distortion that can confuse the user, such as occurs in fish-eye lenses. It is difficult to aim an optic with fish-eye performance. This is because the field is not flat, and angular motion of the telescope will not produce linear motion of the target.

GRAY LEVELS FOR HUMAN OBSERVERS

Subject: Human eye, displays, ergonomics, systems

The Rule

From experiments, we know that 5-bit resolution (or 32 perceived gray levels) is adequate for most applications in which humans attempt to detect targets against a background.

Basis for the Rule

This is based on empirical observations.

Cautions and Useful Range of the Rule

It is applicable to normal, healthy humans viewing standard monochrome (grayscale) CRT displays.

It assumes that the 5-bit resolution gives sufficient signal to noise and signal to background on the screen.

Usefulness of the Rule

The rule is useful for estimating the gray scale required and selecting the balance between dynamic range and sensitivity in systems that use electronic focal planes and displays.

Notes and Explanation

To display sufficient target rendition with 5-bit resolution (32 gray levels), the contrast ratio should be $\geq 19{:}1$. This is because, if we assume that each gray level must have an intensity of 10 percent over the previous, then $\Delta I/I = 0.1$. The contrast ratio can be calculated from:

$$\text{Contrast ratio} = \left(1 + \frac{\Delta I}{I}\right)^{n-1}$$

where ΔI = difference in intensity from one gray level to the next
$\quad I$ = intensity
$\quad n$ = number of gray levels such that the required contrast ratio (given a 10 percent increase between gray levels) is $(1.1)^{31} = 19$

Therefore, the required contrast for 32 gray levels is 19. Conversely, if 64 gray levels are required (16-bit resolution), the contrast ratios must be about $(1.1)^{63} \approx 400$.

Source

I. Spiro and M. Schlessinger. 1989. *Infrared Technology Fundamentals.* New York: Marcel Dekker, 208.

EYE RESOLUTION

Subject: Human eye

The Rule

The human eye can resolve better than one minute of arc and is stabilized by reflex movements.

Basis for the Rule

This rule is based on empirical observations of normal humans.

Cautions and Useful Range of the Rule

As we will see below, the human eye is essentially diffraction limited for those with quality vision or corrective lenses. Another rule in this chapter shows that the pupil diameter is approximately 5 mm for nominal light levels. If we use the standard diffraction formula for defining the angular diameter of a point source,

$$2.44\frac{\lambda}{D}$$

we obtain that, for light of a wavelength of 0.5 microns, the angular resolution is 244 μrad. When converted to arcseconds, we obtain that the resolution is about 45 arcseconds.

Usefulness of the Rule

It is useful in estimation of the resolving power of the human eye.

Notes and Explanation

A person with a visual acuity of 1.5 can resolve 40 seconds of arc, while an average person can resolve 1 minute of arc, which equates a visual acuity of 1. Stated another way,

$$\text{Resolution} = \frac{60 \text{ arcseconds}}{\text{acuity}}$$

Waldman and Wooton[1] make this clear by stating, "The image of the world about us is stabilized on the retina by reflex movements of the eye evoked by stimulation of stretch receptors in the muscles of the head and neck and by hair cells in the vestibule of the ear. Additional eye movements enable us to perceive the world in sharp detail. The eyes are constantly in motion, even when a person is consciously trying to fixate on a given point. Involuntary eye movements are a high-frequency tremor, low speed drift, and flicks–all of low amplitude. These small movements are apparently necessary for vision ... it seems that the rods and cones become desensitized if the irradiance falling on them is absolutely unchanging."

Reference

1. G. Waldman and J. Wooton. 1993. *Electro-Optical Systems Performance Modeling.* Norwood, MA: Artech House, 185.

EYE MOTION DURING THE FORMATION OF AN IMAGE

Subject: Human eye

The Rule

The human eye fixes on particular parts of an image for about 1/3 of a second. Small motions called *saccades* can last as short as a few milliseconds and can involve angular speeds of 1000°/second.

Basis for the Rule

The rule is based in empirical observation of humans.

Cautions and Useful Range of the Rule

The exact size of the saccades and their duration varies somewhat across the human population.

Usefulness of the Rule

Understanding the motion of the human eye is needed for the design of the size, resolution, contrast, and color of displays and their placement (e.g., in a cockpit). This is also useful when designing systems that track or follow the human eye.

Notes and Explanation

Few topics have held the attention of human physiologists as intently as the eye. Vast areas of research have been completed that relate to the function of the eye as a light gathering system, its connection with the muscles of the face, and its connection with the image processing system behind it—the human brain.

During motion, vision is suppressed. A fixation of the eye involves 2 to 3 saccades. The number of eye fixations and sequential pattern describe the search pattern and whether the process is attentive or pre-attentive.

Source

J. Lloyd. 1993. Fundamentals of electro-optical imaging systems analysis. In *Active Electro-Optical Systems*, vol. 4, ed. M. Dudzik, of *The Infrared and Electro-Optical Systems Handbook*, executive ed. J. Accetta and D. Shumaker, Ann Arbor, MI: ERIM, and Bellingham, WA: SPIE, 107.

COMFORT IN VIEWING DISPLAYS

Subject: Human eye, ergonomics

The Rule

The worst usable update rate for a display is 10 Hz, and not much comfort is gained by updating at a rate above 30 Hz.

Basis for the Rule

The rule is based on ergonomics and experience with various types of readout displays.

Cautions and Useful Range of the Rule

This rule is valid for any display viewed by humans. The special irritating effects of a 10 Hz update are somewhat diminished if the display is operating faster (say, 30 Hz) even if the scene is changing at only 10 Hz.

Usefulness of the Rule

This rule is useful for system trade-offs and analysis to determine update rates and in determining the design limits for a display. This is particularly important for displays that will be watched for a long time and for which the proper detection of targets or target motion is critical, such as in military and air traffic control applications.

Notes and Explanation

Most humans are irritated and fatigued easily by a display that is updated near 10 Hz, owing to the human eye-brain processing. An annoying "flicker" is sensed. Interestingly, flames from a fire tend to produce flicker at 8 to 12 Hz, and our annoyance may be an evolutionary adaptation for fire detection. At faster updates (up to about 30 Hz), the eye-brain integrates one frame to the next and smooths out any motion. At slower rates, the eye-brain separates each frame as an individual picture. When frame updates are slow enough (e.g., once per second), the eye-brain assumes that each is an independent picture (as if viewing paintings in an art gallery) and no smoothing of motion occurs. Update rates should be selected that either allow the brain to easily form motion (>20 Hz) or prevent the brain from attempting to form motion (<2 Hz).

Lasers

Revolutions in optics are infrequent. Most of this discipline's history has involved slow evolution in understanding and technological improvements. Lasers, however, have caused a revolution in several applications. First, they have provided impetus for advancements in many areas of EO, since they offer unique diagnostic capabilities essential for producing high-quality systems. EO systems provide a high-quality, stable alignment reference source and allow lens and mirror quality testing using laser interferometry. Second, they have led to the development of a number of applications that would be unthinkable using conventional light sources. Fields such as optical communications and active tracking would be impossible without the unique features of lasers. The spectral purity and compactness of lasers have allowed a number of medical advancements, including those used in eye surgery, elimination of damaged tissue (e.g., gall bladders), and cleaning of clogged arteries.

Finally, lasers, along with television, have been one of the few EO advances to become a part of the lexicon of the average citizen. The invention of lasers in the early 1960s led immediately to the idea of "death ray beams" such as Buck Rogers used. Their enormous brightness and spectral purity have changed many parts of the electro-optics environment and have provided, after about 30 years of development, new advancements in consumer electronics such as CDs, laser printers, and high-performance semiconductors that can be created only with high-performance laser lithography.

As a result of lasers' widespread application, researchers have invested heavily in understanding the characteristics of the beams they produce and the interaction of those beams with various types of targets and detectors. A full understanding of the application of lasers requires new insight into the way electromagnetic waves propagate in the atmosphere. The close relationship of laser light propagation and the medium in which they travel requires this chapter to include a mixture of rules. As a result, the reader will find rules pertaining to the properties of the beams as they might propagate in a vacuum, as well as the way they interact with the propagation medium.

The laser field also has been the source of many interesting stories about how science and industry do business. For instance, when the first optical laser was developed, a press conference was held, pictures were taken, stories were written, and predictions were made. Then, researchers at other laboratories attempted to reproduce the results, using the photos in the newspaper, which conveniently included a ruler so the scale of the objects could be determined. Try as they might, the copycats could not make their versions work. Finally, they approached the researchers who were successful and asked what trick they had done to make their laser work properly. The answer was simple; they never did make the laser shown in the picture work, because the rod was too big to get sufficiently clean. The real laser was much smaller and was not pictured because the press said that it was too small for photographs.

The interested reader has access to a wide variety of texts that describe both the physics and applications of lasers. Siegman provides the current standard of excellence for a sophisticated presentation of laser physics and design approaches. It is probably beyond the skills of any but the most advanced readers, but it is a great resource and should be accessible to everyone who uses or expects to use lasers. Less complex texts are also available in most college books stores. For the entry-level student, the laser industry can be a great resource. For example, the optical manufacturer Melles Griot includes a great deal of useful information in its product catalog. In addition, new laser users will want to look at various magazines such as *Laser Focus World* because they complement the much more complex presentations found in journals such as *Applied Optics* and *IEEE Quantum Electronics*. Optics books should not be overlooked, because most, like the Hecht books mentioned in the bibliography, provide a pretty thorough description of laser operation and applications and provide additional references for consideration.

LASER BRIGHTNESS

Subject: Lasers, systems

The Rule

The brightness of a single-mode laser can be closely estimated by dividing the power-area product by the wavelength squared.

$$B = \frac{PA}{\lambda^2}$$

where B = brightness of the beam (watts/steradian)
P = power of the laser (watts)
A = area of the radiating aperture
λ = wavelength

Basis for the Rule

This rule derives directly from the definition that the on-axis irradiance of a laser is defined as the brightness divided by the range squared. The logic is as follows.

Brightness is defined as the ratio of the power output to the solid angle into which the beam is projected. Therefore,

$$\text{Brightness} = \frac{\text{Power of the beam}}{\text{Solid angle of the beam}}$$

The solid angle of the beam is the area of the beam at the target, divided by the distance to the target squared.

$$\text{Solid angle} = \Omega = \frac{\text{Area}}{R^2}$$

The area of the beam is approximately the square of the product of the beam angle, $\approx\lambda/D$, times the range, or

$$\left(\frac{\lambda}{D}R\right)^2$$

Therefore, the brightness is

$$\frac{P D^2}{\lambda^2} = \frac{P A}{\lambda^2}$$

Cautions and Useful Range of the Rule

The irradiance that is created also depends on the optical quality of the beam. This is sometimes more fully expressed[1] as

$$\text{Irradiance} = \frac{P A}{\lambda^2 BQ^2 R^2}$$

where *BQ* represents the beam quality, measured as a factor that is equal to unity for diffraction-limited performance and a number exceeding unity for all other cases. Using this formulation, we find that brightness is defined as

$$\frac{P\,A}{\lambda^2 BQ^2}$$

Usefulness of the Rule

The rule is useful for making quick estimates of the brightness of a laser to estimate damage effects or human safety. It is also useful for quick estimates of the radiance of a laser to calculate signal-to-noise ratios and so forth.

Notes and Explanations

This rule is related to the antenna theorem,[2] which states that $A\Omega \approx \lambda^2$ can be illustrated simply by noting that, for a diffraction limited beam, the solid angle that is obtained (Ω) is

$$\pi\left(\frac{1.22\lambda}{D}\right)^2$$

and the area of the aperture is such that the product of these two terms is

$$3.67\,\lambda^2$$

Example

A 1 mW HeNe laser with an aperture of 1 mm will have a brightness of about 2 million watts per steradian (W/sr). In contrast, a 1000 W quartz-halogen lamp emits roughly 10 percent of its input power into visible light and over a 4π solid angle. Thus, it has a brightness of about 8 W/sr. Moreover, the laser puts all of its brightness into a very narrow spectral band, whereas the lamp produces light over a wide range of wavelengths. This means that the spectral brightness of the laser is extremely high.

Additionally, a HeNe laser with a 1-mm beam producing at 1 mW radiates as if it were a blackbody of temperature 4.7×10^9 K.[3]

References

1. G. Golnik. 1993. Directed energy systems. In *Emerging Systems and Technologies*, vol. 8, ed. S. Robinson, of *The Infrared and Electro-Optical Systems Handbook*, executive ed. J. Accetta and D. Shumaker, Ann Arbor, MI: ERIM, and Bellingham, WA: SPIE, 451.
2. A. Siegman. 1986. *Lasers*. Mill Valley, CA: University Science Books, 672.
3. G. Fowles. 1975. *Introduction to Modern Optics*. New York: Dover, 223.

APERTURE SIZE FOR LASER BEAMS

Subject: Lasers, optics, systems

The Rule

When a Gaussian beam encounters a circular aperture, the fraction of the power passing through is equal to

$$1 - \exp\left(\frac{-2a^2}{w^2}\right)$$

where a = radius of the aperture

$\quad$ w = radial distance from the beam's center to the point where the beam intensity is 0.135 of the intensity at the center of the beam

Basis for the Rule

The beam intensity as a function of radius is

$$I(r) = I_o \exp\left(\frac{-2r^2}{w^2}\right)$$

where w is defined as above and r is the radius at some point in the beam. The power as a function of size of the aperture is computed from

$$\frac{P(a)}{P_o} = \int_0^a I(r)\, \frac{2\pi r\, dr}{P_o} = 1 - \exp\left(\frac{-2a^2}{w^2}\right)$$

The 0.135 comes from fact that at the $1/e^2$ point of the beam, the intensity is down to 0.135 of the intensity of the center of the beam. Thus, we see that an aperture of $3w$ transmits 99 percent of the beam.

Cautions and Useful Range of the Rule

This rule applies to beams that are characterized as Gaussian in radial intensity pattern. While this is nearly true of aberration-free beams produced by lasers, there are some minor approximations that must be accommodated for real beams.

Usefulness of the Rule

This rule is particularly nice for estimating the size of an aperture needed to pass an appropriate part of a beam.

Notes and Explanation

Of course, as in any system in which an electromagnetic wave encounters an aperture, diffraction will occur. The result is that in the far field of the aperture, one can expect to see fringes, rings, and other artifacts of diffraction superimposed on the geometric optics result of a Gaussian beam with the edges clipped off.

Sources

H. Weichel. 1988. *Laser System Design*, SPIE Class Notes. Bellingham, WA: SPIE, 38.
A. Siegman. 1986. *Lasers*. Mill Valley, CA: University Science Books, 666.

LASER EFFICIENCY, OR "TED MAIMAN'S REVENGE"

Subject: Lasers, systems, management

The Rule

The most inefficient machine on earth, or above it, is a high-power laser. When you put one in a system for any reason, including communications, you have signed up for power and thermal control problems. Often, a passive sensor system can be conjured that is less risky, more efficient, more reliable, and of lower cost.

Basis for the Rule

A laser may consume hundreds of watts but only provide an overage output beam of a few watts. This observation is the result of painful experience and the performance of a wide variety of lasers of all types.

Cautions and Useful Range of the Rule

This is a prejudiced option by the authors, but the evidence is everywhere that laser efficiencies of a few percentage points are quite good. The exception is the larger, generally stationary lasers such as CO_2, which can have efficiencies close to 40 percent.

Usefulness of the Rule

The rule is useful for underscoring that caution is necessary whenever some engineer utters the word "laser." The decision to perform laser tracking, for example, is fraught with challenges. The power consumption and thermal control challenges in the system will increase. Of course, the larger power consumption will impose a mass penalty and initiate a shower of increases in other areas, such as structural stiffness, volume, cost, complexity, and so forth. Only the very brave will carry on and use a laser tracking system when a passive system can be made to work. In the communications field, equivalent challenges exist. In both cases, careful trade-off studies must be performed to ensure that the negative impacts of the inclusion of the laser will be offset by some significant increases in performance.

Notes and Explanation

Ted Maiman invented the laser. Since then, it has been embraced by electro-optical professionals. Although the implementation of a laser may solve serious technical challenges, it often causes equally serious concerns over reliability, power consumption, volume, weight, and cooling.

Of course, as shown in other rules, the high directionality of the laser makes it ideal for applications such as communications, even though it produces waste heat in large amounts.

LED VS. LASER RELIABILITY

Subject: Lasers, systems, optical communication, fiber optics

The Rules

Mean time between failure (MTBF) is around 10^6 to 10^7 hours for LEDs operating at 25° C.

By comparison, commercially available lasers have an MTBF of about 10^5 hours at 22° C.

Basis for the Rules

The rules are based on empirical observations of the current state of the art.

Cautions and Useful Range of the Rules

Advancements in materials technology will continue to improve these values but, for now, these rules apply. Also, be aware that the operating temperature of various lasers has a dramatic impact on the life of these components. Additionally, the MTBF of high-tech high power lasers is usually several orders of magnitude lower.

Usefulness of the Rule

These rules are useful in estimating the reliability of various systems and deriving the cost of the systems.

Notes and Explanation

For optical communication, commercially available lasers have failures from defects in the active region, facet damage, and non-radiative recombination in the active region. Much work is progressing into extending laser reliability.

Source

N. Lewis and M. Miller. 1993. Fiber optic systems. In *Active Electro-Optical Systems*, vol. 6, ed. C. Fox, of *The Infrared and Electro-Optical Systems Handbook*, executive ed. J. Accetta and D. Shumaker, Ann Arbor, MI: ERIM, and Bellingham, WA: SPIE, 258–259.

CAVITY LENGTH VS. LASER OPERATING FREQUENCY

Subject: Lasers, cavity design

The Rule

A change in cavity length produces a change in resonant frequency approximated by:

$$\Delta v \approx -v_n(\Delta L/L)$$

where Δv = change in lasing frequency
v_n = original lasing frequency
ΔL = change in laser cavity length
L = original laser cavity length

Basis for the Rule

Laser physics shows that the round-trip time for a cavity of dimension L must satisfy phase conditions so that amplification can occur. Waves propagating in the cavity that do not match the phase requirements will not be amplified and will be suppressed by ever-present cavity losses. Increasing the cavity length actually reduces the frequency at which the phase condition is satisfied, which is equivalent to saying that long wavelength waves will then be preferentially amplified.

Cautions and Useful Range of the Rule

The change in length must be small (less than, say, 1 percent) compared to the total cavity length.

The modulation bandwidth is limited to the spacing between modes for a multimode laser.

Usefulness of the Rule

It allows quick estimation of potential lasing frequencies.

It allows quick determination of how much change in a cavity is needed to produce a change in frequency or wavelength.

Notes and Explanation

The oscillation (or lasing) frequency of a laser is adjustable over a narrow range by changes in the length of the cavity. Often, one of the cavity end mirrors is mounted on a piezoelectric or magnetostrictive material that allows a small change in length. This change in cavity can be related to the resultant change in frequency by the above equation.

Since frequency and wavelength are directly proportional, the change in wavelength may be found by substituting the original wavelength for the frequency, except that, in the case of wavelength, the minus sign disappears.

Sources

W. Pratt. 1969. *Laser Communication Systems.* New York: John Wiley & Sons, 67.
A. Siegman. 1986. *Lasers.* Mill Valley, CA:. University Science Books, 41.

ON-AXIS INTENSITY OF A BEAM

Subject: Lasers

The Rule

For a beam with no aberrations, the on-axis intensity, in watts per area, is[1]

$$\frac{PA}{R^2 \lambda^2}$$

where R = range
P = beam power in watts
A = transmitting telescope area
λ = wavelength

Basis for the Rule

If aberrations exist and are characterized by beam quality BQ, defined elsewhere in this chapter, then we get

$$\frac{PA}{BQ^2 R^2 \lambda^2}$$

This form is a direct result of the definition of beam quality as being a constant that multiplies the beam spread associated with diffraction. As a result of the multiplication, the beam is bigger in each dimension by a factor of BQ, so the area over which the beam is spread is proportional to $1/BQ^2$.

Cautions and Useful Range of the Rule

This rule is derived from basic laser theory and applies in general. It is limited by the assumption that the beam is propagating without significant atmospheric or other path effects. The discussion below illustrates how that case complicates things.

Usefulness of the Rule

In many applications, the size of the detector is smaller than the beam at the destination. As a result, the on-axis intensity represents the maximum power that can be delivered into such a detector. Of course, as noted in more detail below, the formula above is the ideal. The presence of aberrations in the beam expander and/or laser, along with atmospheric influences, will reduce the power that can be delivered.

Notes and Explanation

The far-field intensity for a circular aperture with reductions due to diffraction, transmission loss, and jitter is[1]

$$I_{ff} = \frac{I_o TK \exp(-\sigma^2)}{1 + (1.57\sigma_{jit} D/\lambda)^2}$$

where I_o = intensity at the laser aperture

T = product of the transmissions of the m optical components in the telescope

K = an aperture shape factor found in Holmes and Avizonis[2] and is found to be very nearly unity in most cases

k = propagation constant = $2\pi/\lambda$

$\Delta\phi$ = wavefront error

σ_{jit} = two-axis RMS jitter

D = aperture diameter

λ = wavelength

$\sigma = k\Delta\phi$ and is the wavefront error expressed in radians

This leads to the following definition of brightness of a laser with jitter and wavefront error:[3]

$$\text{Brightness} = \frac{\pi D^2 PTK \exp(-\sigma^2)}{4\lambda^2 [1 + (1.57\sigma_{jit}D/\lambda)^2]}$$

where P = laser power

We also note that, when the wavefront error, σ, is zero and the jitter term is zero, we get the following:

$$\text{Brightness} = \frac{APTK}{\lambda^2} \approx \frac{PA}{\lambda^2}$$

so the on-axis intensity is equal to

$$\frac{\text{Brightness}}{\text{Range}^2}$$

References

1. K. Gilbert et al. 1993. Aerodynamic effects. In *Atmospheric Propagation of Radiation*, vol. 2, ed. C. Fox, of *The Infrared and Electro-Optical Systems Handbook*, executive ed. J. Accetta and D. Shumaker, Ann Arbor, MI: ERIM, and Bellingham, WA: SPIE, 256.
2. D. Holmes and P. Avizonis. 1976. Approximate optical system model. *Applied Optics* 15(4), 1075.
3. Robert Tyson. 1991. *Principles of Adaptive Optics*. San Diego: Academic Press, 16.

Additional Source

E. Friedman. 1993. On-axis irradiance for obscured rectangular apertures. *Applied Optics* 31(1).

LASER BEAM DIVERGENCE

Subject: Lasers, optics, systems

The Rule

A laser beam's full divergence angle is approximately the wavelength divided by the diameter of the transmitter aperture, or

$$\theta \approx \frac{\lambda}{d}$$

where θ = transmitter beam width (full angle) in radians
λ = wavelength
d = aperture diameter in the same units as the wavelength

Basis for the Rule

This rule derives directly from diffraction theory, particularly as applied to the subject of laser resonators.

Cautions and Useful Range of the Rule

This is a basic result derived from diffraction theory and does not include any of the additional aberrations that occur in real systems. However, this rule is widely used to estimate the size of a laser beam that has propagated through a vacuum and is also frequently used as a first estimate even in atmospheric applications. The rule works fine in environments in which scattering is small in comparison with absorption since, in those cases, the beam shape is not affected.

Usefulness of the Rule

This rule provides quick estimations of minimum beam divergence. Other rules in this chapter provide more detailed procedures that may be applied if more accuracy is required.

Notes and Explanation

If a beam is large in relation to the receiving aperture, the power across the receiving aperture is nearly constant. The power at the receiving aperture is then a function of its placement along the far-field pattern, which can be described using a Bessel function. If the Bessel function is solved for the half-power points, the argument of the Bessel function $\pi d\theta/2\lambda$ is equal to 1.62, and the beam divergence is equal to $1.03\,\lambda/d$.

Note that the divergence is much smaller than the angular distance between the first zeros of the diffraction pattern (the Airy disk), which is $2.44\,(\lambda/D)$.

Sources

A. Siegman. 1986. *Lasers.* Mill Valley, CA: University Science Books, 56.
W. Pratt. 1969. *Laser Communication Systems.* New York: John Wiley & Sons, 5–7.

LASER BEAM SPREAD COMPARED WITH DIFFRACTION

Subject: Lasers, optics

The Rule

A Gaussian spherical wave spreads considerably less than a plane wave diffracted by a circular aperture.

Basis for the Rule

The plane wave has an angular diameter due to diffraction of 2.44 (λ/d) to contain 84 percent of the beam power, whereas a Gaussian spherical wave contains 86 percent of its power in an angular diameter of 2 (λ/d), where d is defined as πw_0, and w_0 is the radius of the waist (or smallest) size of the beam in the beam-forming optics.

Cautions and Useful Range of the Rule

This rule applies for diffraction-limited optics and does not include other aberrations. Furthermore, it does not include the effects of atmospheric scatter in applications in other than within a vacuum.

Usefulness of the Rule

This result should not be overlooked in any application where laser beams are to be propagated. That is, one should not assume that the typical diffraction formula, which defines that beam divergence angular diameter is 2.44 (λ/d), applies for lasers.

Notes and Explanation

The specific results given above derive from a definition of the "size" of the beam. Since a Gaussian beam has an extent that is not well defined, some latitude must be accepted in the power numbers that are selected. For example, Siegman[1] points out examples in which the results vary, depending on how the beam radius is defined.

Reference

1. A. Siegman. 1986. *Lasers.* Mill Valley, CA: University Science Books, 672.

Additional Source

H. Weichel. 1988. *Laser System Design.* SPIE course notes. Bellingham, WA: SPIE, 72.

DIFFRACTION IS PROPORTIONAL TO PERIMETER

Subject: Lasers

The Rule

Clark et al.[1] have shown that the far-field diffraction of a uniformly illuminated aperture of arbitrary cross section is proportional to the perimeter of the aperture.

Basis for the Rule

A rather involved calculation, using diffraction theory, was employed to generate this result.

Cautions and Useful Range of the Rule

This rule is only an approximation but works well. In addition, it applies only in the far field and only in the region near the beam axis. It also assumes that the illumination of the aperture is uniform.

Usefulness of the Rule

This rule provides a quick estimates of diffraction in the far field for beams produced by other than the standard circular and square apertures.

Notes and Explanation

Specifically, the rule states that the beam angle for which half of the power is included is

$$\frac{p\lambda}{\pi^2 A}$$

where p = perimeter of the aperture
λ = wavelength
A = aperture area

Generally, it is quite difficult to compute the far-field radiation pattern from an arbitrary aperture shape. This handy rule simplifies the process and provides answers of adequate accuracy for most purposes. For example, it can be used to show the incremental effect of additional secondary mirror struts or other structural elements that might be inserted into the aperture.

Reference

1. P. Clark et al. 1984. Asymptotic approximation to the encircle energy function for arbitrary aperture shapes. *Applied Optics* 23(1), 353.

GAUSSIAN BEAM SIZE

Subject: Lasers, optics, systems

The Rule

In the far field, the full beam width of a Gaussian beam can be approximated by:

$$\varphi = \frac{2\lambda}{\pi\omega_o}$$

where φ = transmitter full-angle beam width in radians
  λ = optical wavelength in meters
  π = 3.1415
  ω_0 = Gaussian beam waist radius in meters

Basis for the Rule

This definition of beam width is based on the location of the $1/e$ points in the beam electric field. This is one of several common ways that beam spread is defined for Gaussian laser beams. Siegman[1] describes several of them, including the more conservative 99 percent criterion in which the size of the beam is defined as the area that includes 99 percent of the energy in the beam.

Cautions and Useful Range of the Rule

As pointed out above, rules of this type must be interpreted within the context of the definition of what constitutes the beam. Furthermore, the rule only applies in the far field, which is defined as range at which the range is greater than

$$\frac{\pi w_o^2}{\lambda}$$

This last distance is called the *Rayleigh range* and is equal to the distance from the waist at which the diverging beam is the same size as the waist.

Usefulness of the Rule

It is useful in determining the far-field pattern of a laser.

Notes and Explanation

The beam width of a Gaussian beam is defined as the full width across the beam measured to the e^{-2} irradiance levels. One often encounters the beam width defined at the full width half maximum (FWHM) points. To convert a Gaussian beam profile specified at FWHM to the equivalent $1/e^2$ points, multiply it by ≈ 1.17.

Reference

1. A. Siegman. 1986. *Lasers*. Mill Valley, CA.: University Science Books, 671.

Additional Source

G. Kamerman. 1993. Laser radar. In *Active Electro-Optical Systems*, vol. 6, ed. C. Fox, of *The Infrared and Electro-Optical Systems Handbook*, executive ed. J. Accetta and D. Shumaker, Ann Arbor, MI: ERIM, and Bellingham, WA: SPIE, 15–16.

CROSS SECTION OF A RETRO-REFLECTOR

Subject: Lasers, optics, phenomenology, systems

The Rule

The laser radar cross section of a cube corner retro-reflector exposed to an approximately plane wave and viewed from the position of the laser is

$$\frac{D^4}{4\lambda^2}$$

where D = edge dimension of the cube
λ = wavelength

Basis for the Rule

The diffraction limited return beam angle is $1.22(\lambda/D)$, so the beam fills an area of $\pi[1.22(\lambda/D)]^2$ for a circular retro. This is the same as the solid angle of the beam emitted by the retro and is approximately to $4(\lambda^2/\text{area})$. This is also the exact result for a retro with a square aperture. Cross section is defined as the inverse of the solid angle divided by the area. Thus,

$$\text{Cross-section} = \left(4\frac{\lambda^2}{\text{area}^2}\right)^{-1} = \frac{D^4}{4\lambda^2}$$

Cautions and Useful Range of the Rule

This rule applies to *cube corner* or *corner cube* retro-reflectors only. Occasionally, such devices will have a slightly smaller cross section due to less than unity reflection and less than perfect tolerances on the angles of the mirrors.

Usefulness of the Rule

This rule provides an immediate estimate of the detectability of an object equipped with a retro-reflector. It also allows the reflector to be sized so that detection at an appropriate range and for a particular laser power can be estimated. Additionally, targets (even non-cooperative ones) frequently have structures that approximate cube corners, giving them a much larger signature than would otherwise be assumed.

Notes and Explanation

Retro-reflectors are often used in tracking and pointing experiments to ensure that the target is detected and the experiments can be carried out reliably. The reason is clear. By providing even the smallest retro-reflector, the target's signature is large and can be seen at great distances, even with laser systems of modest power.

An unfortunate term has crept into the literature concerning cube corners. The expression *corner cube* is much more common but is misleading. A retro-reflector can be formed from the corner of a cube over restricted angles. The term *corner cube* does not describe the geometry of any retro-reflector currently in use, but the term is frequently used.

Example

As an example, consider a cube of 0.1 m edge dimension exposed to visible $(0.5\,\mu)$ light. The result is 100,000,000 m^2, 100 km^2, which is rather huge. Thus, the presence of a retro-reflector makes a target behave as if it is many orders of magnitude larger than it actually is.

BEAM QUALITY

Subject: Lasers, optics, systems

The Rule

Beam quality is defined as

$$BQ = \exp\left[\frac{1}{2}(2\pi\ WFE)^2\right] = \frac{1}{\sqrt{S}}$$

where BQ = (unitless)
$\quad S$ = Strehl ratio (unitless)
$\quad WFE$ = wavefront error expressed in waves

Basis for the Rule

The best beam focusing and collimation that can be obtained is derived from the diffraction theory for plane and Gaussian beams encountering sharp-edged apertures, as described in virtually every optics and laser book. In those analyses, it is assumed that the wavefront is ideal and that there are no tilt or higher-order aberrations in the phase front. This parameter has been developed to deal simply with the additional impact of nonuniform phase fronts in those beams. In a wide variety of applications, this definition is used to define the additional spreading that will be encountered in focused or parallel beams.

Cautions and Useful Range of the Rule

There are many other definitions of beam quality, so the reader is cautioned to understand what is meant by BQ in each particular application.

These characterizations of a laser beam are effective measures when the beam is nearly diffraction limited. For highly aberrated beams, it may be difficult to establish the beam quality. For example, using the definition that relates beam quality to beam size, a highly aberrated beam will have an ill-defined diameter that varies with azimuthal angle, thus limiting the usefulness of the definition.

Usefulness of the Rule

This rule provides a simple parameter that defines the beam spread of a laser beam. For example, the dimension of the spot of a beam will be expressed as $(\lambda/D)\,BQ$, rather than the ideal (λ/D). This means that the spot will cover an area that is proportional to BQ^2. Therefore, we find that the energy density in the beam will depend on $1/BQ^2$.

Notes and Explanation

This parameter defines a shortcut way to characterize how much a beam deviates from the ideal, diffraction-limited case. As mentioned above, this can be an effective definition if the BQ is close to unity. For very poor beam quality, such as might result from turbulence and other atmospheric effects, the entire concept of a well defined beam becomes useless, and this definition fails to characterize the beam.

BQ can also be defined in terms of the power inside a circle at the target.

$$BQ = \sqrt{\frac{P_{ideal}}{P_{actual}}}$$

where the powers are compared at a common radius from the center of the target.

The effect of beam quality is included in the typical diffraction spreading of a beam by

$$\theta_D = \left(\frac{2BQ\lambda}{\pi D}\right)$$

where BQ is the beam quality at the aperture. Clearly, when BQ is unity, the diffraction angle is the same as described in another rule covered in this chapter,

$$\theta_D = \left(\frac{2\lambda}{\pi D}\right)$$

Thus, we see that BQ is included as a linear term is estimating the beam spread of a laser.

The Strehl ratio refers to the on-axis intensity of a beam as compared with the ideal value. Evidently, if $BQ = 1$, $S = 1$, and the two are inversely related, with a $BQ > 1$ inducing a rapid decrease in on-axis intensity.

Source

G. Golnik. 1993. Directed energy systems. In *Emerging Systems and Technologies*, vol. 8, ed. S. Robinson, of *The Infrared and Electro-Optical Systems Handbook*, executive ed. J. Accetta and D. Shumaker, Ann Arbor, MI: ERIM, and Bellingham, WA: SPIE, 472.

PULSE BROADENING IN A FABRY-PEROT ETALON

Subject: Lasers, optics, systems

The Rule

The final pulse width (in seconds) of a laser pulse after N trips through a Fabry Perot etalon is given by

$$\approx 3.5 \times 10^{-11} F d \sqrt{N}$$

where F = finesse of the etalon (which can be as large as about 100)
$\quad d$ = thickness in cm
$\quad N$ = number of trips

Basis for the Rule

Siegman[1] provides an exercise for those using his book in which he includes a more complex representation of the formula and a challenge to come up with the simpler form presented above.

Cautions and Useful Range of the Rule

The result applies to the pulse width after many trips through the etalon. Curiously, the more complex form includes the length of the initial pulse. This version does not.

Usefulness of the Rule

The rule provides a simple method for estimating the length of a pulse that has passed through an etalon many times. Note that the pulse continues to grow without bound, since N is present under a square root.

Notes and Explanation

Etalons provide extremely high-performance optical filters in many EO applications. In typical applications, the etalon is able to resolve wavelengths down to hundredths of a nanometer or smaller. In addition, etalons form the basis of most laser resonators and can be used to control the number of modes that propagate freely in the resonator.

Reference

1. A. Siegman. 1986. *Lasers*. Mill Valley, CA.: University Science Books, 360.

POINTING OF A BEAM OF LIGHT

Subject: Lasers, servos, controls, and systems

The Rule

The probability, P_h, that a beam of angular radius θ_d will be pointed to within its radius (θ_d) is

$$\sigma = \frac{\theta_d}{2} \frac{1}{\sqrt{-2 \ln (1 - P_h)}}$$

Where the pointing error is defined by a Gaussian with a standard deviation of σ.

Note that the definition of the beam angular radius is a choice to be made by the user of this equation. For example, it could be the half-power point of a Gaussian laser beam or the $1/e$ point. The equation applies in either case, but the choice must take into account that the definition of beam radius determines the actual amount of power imposed on the target. Figure 9.1 illustrates the example. Note that the equation applies, regardless of the distribution of energy in the beam. For example, if the beam has a Gaussian distribution of energy, θ_d could describe the $1/e$ points of the beam.

Basis for the Rule

The probability that the beam is within a solid angle Ω is

$$P(\Omega) = \frac{\exp\left(-\dfrac{\theta^2}{2\sigma^2}\right)}{2\pi\sigma^2}$$

The probability that the beam is pointed at the target within the angle θ is

$$P(\theta) = P(\Omega)\frac{d\Omega}{d\theta} = \frac{\exp\left(-\dfrac{\theta^2}{2\sigma^2}\right)}{\sigma^2}$$

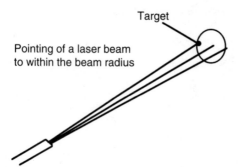

Target

Pointing of a laser beam
to within the beam radius

Figure 9.1 Example of Beam Pointing

If we want to compute the probability that the beam is pointed to within one-half of the beam diameter, we integrate $P(\theta)$ over that interval;

$$P = \int_0^{\theta_d/2} \theta \frac{\exp\left(-\dfrac{\theta^2}{2\sigma^2}\right)}{\sigma^2} d\theta$$

Of course, other measures of merit can be chosen as well, such as $1/10$ of the beam diameter. For the case of $\theta_d/2$, some manipulation results in

$$P = 1 - \exp\left(-\frac{\theta_d^2}{8\sigma^2}\right)$$

Solving for σ, we get the equation at the start of this rule.

Cautions and Useful Range of the Rule

The rule, as stated, applies to the illumination of a point target by a beam. Larger targets are, of course, easier to hit. The size of the target is added to $\theta_d/2$. The analysis also assumes that there is no bias in the pointing of the beam.

Usefulness of the Rule

This rule can be quite useful in defining the pointing necessary to illuminate a target or the sizing of the beam necessary to achieve the desired probability of hit.

Notes and Explanation

If one does calculations at the half-power point or $1/e$ points of a laser beam's intensity, one is building in a margin. Usually, the power will be 50 to 70 percent higher. If multiple "hits" or observations are allowed, then there is a comfortable margin built in, as it is unlikely that random errors will result in several observations at the minimum points.

Example

Suppose we have a beam that has a divergence, measured in half-cone angle, of 10 mrad. What pointing is necessary to ensure that the beam will encounter a point target with a probability of 0.99? This works out to be 1.67E–3 radians. This result is consistent with the general conclusion that for high probability of pointing to within the beam radius, the pointing must be about $1/5$ of the radius.

Of course, the equation can be manipulated to calculate the probability of hit given the beam dimension.

LIDAR PERFORMANCE

Subject: Lasers, LIDARs

The Rule

A laser ranger, also called a LADAR or a LIDAR, has a signal-to-noise ratio that varies as $1/R^2$ when the beam is smaller than the target at the target, and as $1/R^4$ when the beam is bigger than the target.

Basis for the Rule

The geometry of the problem shows that, when the beam is smaller than the target, all of the radiation from the laser hits the target. The reflected light is scattered into a hemisphere, a part of which includes the receiver. Since the scattered light emanates approximately from a point, the amount of light received is proportional to $1/R^2$

Cautions and Useful Range of the Rule

Atmospheric effects and the surface properties of the target add additional effects to the expected performance of a LIDAR. For example, a target with a specular surface can actually make the target nearly invisible because the reflected radiation may have very little component in the direction of the receiver. In addition, highly absorptive surfaces will further suppress the amount of reflected energy that is detected. Thus, flat, highly absorptive surfaces may cause the range sensitivity of a LIDAR to be far worse than $1/R^4$. Again, the $1/R^2$ applies only when the footprint of the projected beam completely falls on the target.

Usefulness of the Rule

This rule will remind us that the large beam solution in a laser radar is no solution at all since the signal will decrease as the fourth power of the range (or faster). In most cases, this will be a fatal flaw in the design. Additionally, this rule is a reminder that for some cases, the signal follows a $1/R^2$ relationship rather than $1/R^4$ law that is most common for laser detection systems.

Notes and Explanation

Figure 9.2 shows how the beam divergence affects the performance of a ranging system. The small circle represents a laser beam of small divergence that encounters the target. In this case, none of the light from the laser is lost, and the target is fully encountered. The larger beam does not use all of the available light. A substantial part of the light is lost and cannot contribute to the signal. It should be noted that, in assessing the relative performance of the two cases, we assume that both have the same amount of laser power available.

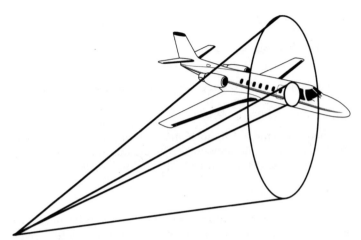

Figure 9.2 Effect of Beam Divergence

LASER BEAM WANDER VARIANCE

Subject: Lasers, atmospherics, systems

The Rule

The variance (σ^2) in the position of a beam propagating in the atmosphere is

$$1.83 \ C_n^2 \ \lambda^{-1/6} \ L^{17/6}$$

where L is the path length, and the path is horizontal. The square root of this number is the standard deviation in the beam wander. The distance the beam travels is L, and C_n^2 is the atmospheric structure constant. All units are usually expressed in meters.

Basis for the Rule

A whole generation of atmospheric scientists has worked on the problem of laser beam propagation in the atmosphere. Ultimately, all of the work derives from seminal analyses performed by the Russians Rytov and Kolmogorov. Fried has also made important contributions to the theory. Military and astronomical scientists have extended the theory and made considerable progress in demonstrating agreement between theory and experiment. The theory is too complex to repeat here. Fortunately, the effect on propagation can be expressed with relatively simple algebraic expressions such as the one shown above.

Cautions and Useful Range of the Rule

As with any rule related to the atmosphere, the details of the conditions really determine the propagation that will be observed. This result assumes that the value of C_n^2 along the path is constant and is of such a value that the turbulence effect falls into the category of "weak." This means that the variance in the beam intensity is less than about 0.54. Otherwise, the assumptions inherent in Kolmogorov's adaptation of Rytov's work no longer apply, and the results are flawed.

Usefulness of the Rule

Use of this rule defines the size that a receiver must have to encounter the bulk of a beam used for communications, tracking, or other pointing-sensitive applications.

Notes and Explanation

The mathematics behind this analysis, which was first done by Tatarski, is beyond the scope of this book. Suffice it to say that the result shown above is a rather substantial simplification of the real analysis that must be performed. For example, Wolfe and Zissis[1] provide a more complete analysis and show how the beam wander is translated into motion of the centroid of the beam in the focal plane of a receiver.

Example

The nighttime value of C_n^2 is about 10^{-14} m$^{-2/3}$. For a path length of 5000 m and a wavelength of 0.5 microns, $\sigma \approx 78$ mm.

Reference

1. W. Wolfe and G. Zissis, eds. 1978. *The Infrared Handbook*. Ann Arbor, MI: ERIM, 6-37.

Additional Source

H. Weichel. 1988. *Laser System Design*, SPIE course notes. Bellingham, WA: SPIE.

LASER CROSS SECTION

Subject: Lasers

The Rule

The laser cross section of a target is about 10 percent of the target area projected toward the laser.

Basis for the Rule

This rule is based on empirical observations, grand generalizations, and some simple calculations, along with typical reflectivity of materials.

Cautions and Useful Range of the Rule

Because this is a grand generalization, extreme caution should be used; however, this estimate is about right most of the time.

   This assumes a surface that is a Lambertian reflector of average reflectivity.

   Try a rigorous calculation of laser cross section before designing a system. However, this rule is handy for quick-thought problems and what-ifs when other data is lacking.

Usefulness of the Rule

It is useful for quick estimates of cross section when a target is being tracked by an active laser tracker.

Notes and Explanation

The effective laser cross section of a target is generally much less than the projected area. For example, consider a spherical target with a diffuse (Lambertian) surface. The projected area of the sphere is πr^2, but the radiant intensity is the incident irradiance times the reflectivity and divided by π. Since the reflectivity of most targets is less than 80 percent and greater than 20 percent, the effective cross section is between $0.2/\pi$ and $0.8/\pi$ times the physical cross section. The first value is 0.06 and the latter value is 0.25. On average, the effective cross section from a radiometric perspective is on the order of 10 percent—more for high cross section targets and less for stealth targets.

PEAK INTENSITY OF A BEAM WITH INTERVENING ATMOSPHERE

Subject: Lasers, phenomenology, atmospherics, environment

The Rule

Peak intensity (W/m²) of a beam going a distance L (in meters) is

$$\frac{P\,e^{-\varepsilon L}}{\pi L^2\,(\sigma_L^2 + \sigma_B^2)}$$

where L = distance

σ_B = effect of blooming, in radians

ε = atmospheric extinction, in meters^{-1}

P = beam power at the transmitter, in watts

σ_L = combined effect of linear beam spread functions and is equal to

$$\sqrt{\sigma_D^2 + \sigma_T^2 + \sigma_J^2}$$

σ_D = combined effect of diffraction and beam quality in radians

σ_T = effect of turbulence, in radians

σ_J = effect of jitter, in radians

Basis for the Rule

The beam intensity at some distance L is the result of the combined effect of beam spreading and atmospheric attenuation. The latter is contained in the exponential term in the numerator. It contains both the absorption, which removes energy from the beam, and scattering, which redirects the energy but removes it from the beam. The denominator simply describes the area over which the beam will be spread at the distance L. It relies on several terms to describe the size of the beam, as described above. The rule "On-Axis Intensity of a Beam" (page 161), provides more information on the basis of this rule.

Cautions and Useful Range of the Rule

Of course, the description of the beam shape is a simplification, particularly with respect to the atmospheric effects. The range of limitation really applies to the parts of the rule relating to atmospheric effects. The estimation of the impact of the atmosphere applies so long as the turbulence falls into the "light" (as opposed to "heavy") category. A number of rules in Chapter 2, Atmospherics, deal with how to compute the conditions that apply for light turbulence.

Usefulness of the Rule

The rule provides a quick and easy-to-program model of the intensity of a laser beam as a function of atmospheric conditions and the range to the target.

Notes and Explanation

The diffraction effect is

$$\sigma_D^2 = \left(\frac{2BQ\lambda}{\pi D}\right)^2$$

where BQ = beam quality at the aperture (Note that when BQ = 1, we get the diffraction effect, which cannot be avoided.)

When $D/r_0 < 3$, which applies for short paths,

$$\sigma_T^2 = 0.182\left(\frac{\sigma_D}{BQ}\right)^2\left(\frac{D}{r_0}\right)^2$$

where r_0 = Fried's parameter, discussed in Chapter 2, r_0 varies from about 10 cm for a vertical path through the entire atmosphere to several meters for short horizontal paths.

When $D/r_0 > 3$

$$\sigma_T^2 = \left(\frac{\sigma_D}{BQ}\right)^2\left[\left(\frac{D}{r_0}\right)^2 - 1.18\left(\frac{D}{r_0}\right)^{5/3}\right]$$

Also, note that the first equation reverts to

$$\frac{PA}{BQ^2\lambda^2 L^2}$$

when there are no atmospheric effects, which is consistent with the "On-Axis Intensity of a Beam" rule for the case in which no atmosphere is present.

Source

R. Tyson and P. Ulrich. 1993. Adaptive optics. In *Emerging Systems and Technologies*, vol. 8, ed. S. Robinson, of *The Infrared and Electro-Optical Systems Handbook*, executive ed. J. Accetta and D. Shumaker. Ann Arbor, MI: ERIM, and Bellingham, WA: SPIE, 198.

ATMOSPHERIC ABSORPTION OF A 10.6-μm LASER

Subject: Lasers, atmospherics, systems

The Rule

The absorption coefficient (in dB/km) of Beer's law can be approximated for a 10.6-μm laser given the following conditions as:

Clear: $1.084 \times 10^{-5} p (P + 193p) \left(\dfrac{296}{T} \right)^{5.25} + 625 \left(\dfrac{296}{T} \right)^{1.5} \times 10^{-970/T} + \dfrac{1.4}{V}$

Rain: $1.9 R^{0.63}$

Snow: $2 S^{0.75}$

Fog: $\dfrac{1.7}{V^{1.5}}$

Dust: $\dfrac{5}{V}$

where V = visibility in km
S = snowfall rate in mm/hour
R = rainfall rate in mm/hour
T = temperature in K
P = atmospheric pressure in millibars
p = partial pressure of water vapor in millibars

Basis for the Rule

These rules are the result of empirical studies.

Cautions and Useful Range of the Rule

These are estimates for the atmospheric extinction of Beer's law, so they must be used with Beer's law only. Note that the units are dB/km, so a conversion must be employed if the answer is desired in units of transmission percentage or decimal notation.

Clearly, these simple equations cannot do justice to the real behavior of the atmosphere. Furthermore, it must be remembered that the specific numbers used here apply only to the 10.6-μM band; therefore, the frequently performed extrapolation to other bands is only done with extreme caution.

Usefulness of the Rule

This type of rule provides a quick estimate of atmospheric transmission that can be helpful during the planning and execution of field experiments. Atmospheric transmission of any wavelength usually can be obtained in adequate detail by using codes like LOWTRAN and MODTRAN. However, we all desire simple rules that can help us deal with complex issues in an easy way. This rule provides a rough idea of the transmission of the atmosphere in the very important 10.6µm band produced by a CO_2 laser. It provides an estimate for a variety of atmospheric conditions.

Notes and Explanation

Because these results are empirical in nature, there is little to say about the physics that causes them to be true. However, their validity has been borne out in field experiments.

Source

G. Kamerman. 1993. Laser radar. In *Active Electro-Optical Systems*, vol. 6, ed. C. Fox, of *The Infrared and Electro-Optical Systems Handbook*, executive ed. J. Accetta and D. Shumaker, Ann Arbor, MI: ERIM, and Bellingham, WA: SPIE, 26.

10

Management

Any writings about the management of electro-optics must be closely related to that of project management. Most electro-optic (EO) systems tend to follow the classic U.S. Department of Defense (DoD) life cycle.

EO systems have a life of a several decades, starting with a small concept group trying to sell an idea to a customer who will pay for the development. In itself, this phase can last a decade—and rarely less than a year. If successful, the small team grows into a proposal team. Again, if successful, this team grows into a group of visionaries who flow the desires to requirements and specifications for the *system requirements review* (SRR) and mature the wild ideas to something that can stand the scrutiny of a *preliminary design review* (PDR). Following this, a larger team of managers and engineers is usually formed to develop a design to a level where prototypes can be made and tested for the *critical design review* (CDR). Often, following a successful CDR, the design is finalized, the technical "i" dotted and the technical "t" crossed. Production usually starts within a year, with some changes being necessitated by the production process. As the product matures in the field, the logistics support engineers and managers dominate as they attempt to keep it running and doing what the marketeer promised in the first place.

Although a modern term, *project management* is perhaps the most primordial form of management. The history of project management must go back to cavemen who went on collective hunts and focused on a single goal: food. When agriculture was developed, humans had more time to devote to non-food megaprojects such as the pyramids (Yucatan, Asia, and Egypt), earth carvings (Central and South America, Europe) and giant statues (Polynesia and Europe). Certain historic figures such as Napoleon, Gangues Khan, Lao Tzu, and various Roman emperors and Egyptian rulers were great project managers. However, the early days of management were tightly interwoven with religion, government, and/or the military (not much has changed, huh?). Distinct writings separate from these three dogmas were not common until the Renaissance, when merchants were simply trying to make a secular buck.

Frederick Taylor is often called the *father of scientific management* with his popular writings of around 1900. Another early milestone was Henry Ford's assembly line concept (still rarely employed in electro-optics). Shortly afterward, Thomas Edison clashed with some creative free-souls (e.g., Nikola Tesla) because of his repressive "management" ideas about scientific discovery, research, and development.

Then, megascience and megaprojects were created with the Manhattan Project lead by General Groves and Robert Oppeheimer. In modern times, this was the first scientific megaproject. Under the auspices of Groves and Oppeheimer, the diverse personalities of several prima donnas were juggled with budget constraints, security problems, impossible schedules, questionable test results, moral dilemmas, and limited resources—just like any modern electro-optical project. Since then, large-esque engineering and science has been the norm, with projects such as rural electrification, Apollo, laser fusion, space shuttles, CERN, Hubble, human genome mapping, SDI, the supercollider, and now the information superhighway.

However, ever since the early days of largesse, there were always some like Kelly Johnson of Lockheed's *Skunk Works* who bucked the traditional, highly structured organizations. He preferred developing complex systems using a small, dedicated team that was unencumbered with a bureaucratic hindrances. That is how we got the U-2, SR-71, and the F-117 with quick development times and relatively low costs. In a better-late-than-never copycat action, many modern companies are now proceeding to rid themselves of the latent bureaucracy and empower the middle management, scientists, and engineers as Johnson did in the 1960s.

The twentieth century has been sprinkled with fad management techniques, with a few appearing every decade. It is a management-theory jungle out there, with each theory, like a fruit, providing a much needed nutrient. However, poor nutrition will result if only one fruit is eaten with the abandonment of every other food. Perhaps the best advice is that old rule-of-thumb, "If it ain't broke, don't fix it."

For the interested reader, there are always new texts and publications in the area of project management. For background, see the appropriate texts listed in the bibliography.

The following are a collection of light-hearted, tongue-in-cheek rules as they apply to the management and marketing of electro-optical systems. They are general truisms and not accurate all of the time in every case, but they are often useful.

VALUE OF EARLY INVESTMENT

Subject: Management, system development, the sad facts of life

The Rule

Spend the required money up front to do the design right. Do enough testing to find the design weaknesses, or pay more later to correct the problem.

Basis for the Rule

The fact is that fixing design flaws (see Fig. 10.1) always costs more than just doing the design right. It is always cheaper to fix a paper drawing than to recall and retrofit a fix to a production unit. This basic rule results from empirical observations and common knowledge.

Cautions and Useful Range of the Rule

Unfortunately, the enemy to this wisdom is the customer funding profile. Often, the program manager has no choice but to scrimp at the beginning due to funding shortages.

This applies to all electro-optic systems and cars, boats, buildings, airplanes, and so on.

Usefulness of the Rule

It is a reminder that, for every dollar spent providing a good design up front, several dollars are saved in the mature part of the program.

Notes and Explanation

Basically, this concept boils down to "pay me now or pay me late." And, as expected, if you pay me later, you will pay more.

In early design phases of a program, changes can be implemented for little cost. However, the same changes result in dramatic cost increases after the product is made (just ask any car company about recalls). Electro-optics is a design intensive industry with rather small production runs. It is important to do the design right from the beginning. In other words, spend the money and time from the beginning to do a detail design, check it, and test it to wring out any deficiencies and problems before going into production. In these cost-conscious times, this common knowledge is often overlooked.

Source

Rule Supplied by Joe Calabretta, 1995.

Figure 10.1 Design Flaws

THE FINAL PERFORMANCE GENERATES THE MOST COST, OR THE "80–20" RULE

Subject: Management, project management

The Rule

The last 10 or 20 percent of performance generates most of the cost and problems.

Basis for the Rule

This long-lived management tenet results from astute observations of cost and schedule versus performance on many projects. This rule is based on the fact that, as performance is increased, it becomes asymptotically difficult to achieve any extra performance. As a result, during development, most of a system's cost and problems are tied up in achieving these final levels of performance.

Cautions and Useful Range of the Rule

Because this is a great and gross generalization, caution should always be applied.

Usefulness of the Rule

This should always be considered when proposing an effort, during project management, when scheduling, and in the monitoring of projects.

Notes and Explanation

Sometimes this is stated as "80 percent of the cost is attributed to 20 percent of the performance."

It is apparent to anyone who has shopped for a car that the cost of increasing performance increases nonlinearly as performance increases. Hence, Ferraris and Mercedes cost more than Lexuses and Saabs, which cost more than Toyotas and Volkswagens. This trend is present throughout the engineering, materials, and manufacturing worlds. When you find a way to violate this rule, you will have a hot-selling product.

This is why a Mercedes costs more than a Ford.

SINGLE IDEA

Subject: Management, engineering, science, project management, life

The Rule

"Nothing is more dangerous than an idea, when it is the only one that you have."

Basis for the Rule

Astute empirical observations by Mr. Chartier.

Cautions and Useful Range of the Rule

Sometimes the one idea that you have is the best, but don't count on it.

Usefulness of the Rule

It is useful to recall this rule, which underscores a serious caution, often during the proposal, design, and test processes.

Notes and Explanation

Electro-optical system design, production, and test does not lend itself to a singularity. Rather, creative and different perspectives usually provide the most cost-effective solution. The search for such alternative solutions (or ideas) is sometimes stopped after the first successful idea has surfaced.

The above quote is from Emile Chartier. It exemplifies an important fundamental of human nature. A corollary may be found in the old adage, "There is more than one way to skin a cat." Let's take this a little further and apply it to project management of E-O systems. A single idea, concept, design, or procedure, when unchallenged, often leads to failure. It can lead to groupthink and preclude the right solution, which is being promoted by your competition. The lack of multiple paths often leads to a dead end in the design, causing you to abandon the effort, while another research lab is proceeding with a different idea. A single-point design leads to a brittle system that will fail when countermeasures are applied.

Project Management Rule of Ten

Subject: Management, project management

The Rule

If you have ten people working for you, it is a full-time job to let them report to you. If you have more than ten people reporting to you, then you are wasting everyone's time. If you have fewer than ten people reporting to you, then you might get something done.

Basis for the Rule

This rule is based on painful empirical observations by the authors. It also stems from the level of required monitoring of your subordinates that is typical in modern America and Europe.

Cautions and Useful Range of the Rule

This is an average, based on normal E-O project situations.

Usefulness of the Rule

This can be used to estimate the management/project structure and to determine how much you will actually contribute.

Notes and Explanation

Generally, if you have more than seven to ten people directly reporting to you, then you do not have the time to figure out what they are telling you. Conversely, if you have fewer than about seven, you can do some level of independent work yourself. The occasional exception to this rule is if the people are not in competition with each other and are generally isolated from each other. This rarely occurs in science and engineering but frequently does in the retail industry. It has been reported that some retail managers have successfully managed several hundred people (but, remember, they weren't crybaby electro-optic prima donnas).

Pick Any Two

Subject: Management, project management, facts of life

The Rule

A developing system (e.g., at the preliminary design review) can have any two of the following qualities:

- low cost
- high reliability
- high performance
- fast delivery

Basis for the Rule

This is founded on tongue-in-cheek empirical observations.

Cautions and Useful Range of the Rule

Low cost, high reliability, high performance, and fast delivery are all relative (see Fig. 10.2). A million dollars may be considered low cost for a given system at a certain reliability, performance and delivery. Increase the reliability and the cost will increase relative to its initial cost. Likewise with performance and quicker deliveries.

This assumes comparison of the same type of cost, reliability, and so forth. For instance, an increase in reliability may cause a decrease in life cycle cost (but rarely development cost).

The above does not account for revolutionary changes in technology or production techniques. Said another way, it may take a revolution in technology to invalidate these rules.

Finally, this assumes the design was not seriously flawed, so that a minor design change will affect several attributes favorably.

A less mature system can be any three of the above.

Usefulness of the Rule

The rule reminds one of the difficulties of conflicting requirements.

Notes and Explanation

Often, the requirements (or attributes) of an electro-optical project compete with each other and may even be contradictory. For instance, it is usually difficult to increase performance and reliability while reducing cost. One requirement may act like nitroglycerin to another requirement; just moving it can cause the other to explode.

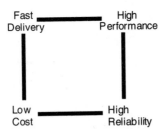

Figure 10.2 Pick Any Two (well, maybe three if you are lucky)

Usually, an astute engineering/manufacturing organization can fuse at least two opposing requirements together and satisfy both simultaneously for a system of mature design. For a developmental system, they can usually promise three with straight faces.

MURPHY'S LAW

Subject: Management, nature, life

The Rule If anything can go wrong, it will.

Basis for the Rule

This rule was developed by Mr. Murphy from painfully acquired experience and has become part of the American lexicon.

Cautions and Useful Range of the Rule

Although a gross generalization, this rule seems to always work.

Usefulness of the Rule

It underscores a basic tenet of nature and reminds us to expect problems and failures.

Notes and Explanation

Murphy was some type of technician/engineer/genius (every discipline seems to claim him as one of theirs) who developed this simple concept, which will probably be forever associated with his name. The rule is true if the time allowed for something to break (or go wrong) is infinity. For the shorter time periods that we humans deal with, the rule may not apply in every case. However, the authors could find no one who could point out an instance in which the rule failed! Maybe the application of the rule to itself is the only case?

MOST OF THE COST IS IN THE BEGINNING, OR THE "90–10" RULE

Subject: Management, project management, system development

The Rule

Ninety percent of the cost of a system is determined in the first ten percent of the design effort.

Basis for the Rule

This assertion is based on empirical observations, and astute observation combined with fundamental knowledge of project management.

Cautions and Useful Range of the Rule

This applies to all electro-optic systems and generally applies to any engineering project. In rare cases (e.g., ones employing many nondevelopmental components), the above distribution might be more balanced.

Usefulness of the Rule

The rule underscores the usefulness of concurrent engineering. It also emphasizes an important fact to keep in mind during proposal and concept definition phases of programs.

Notes and Explanation

Managers, engineers, and designers should be considering the cost of the program from the beginning, as their decisions may affect cost for decades to come. The concept definition and early architecture phases of an electro-optical project frequently determine most of the cost. The system can be designed for low cost or high cost (usually with high performance). The early decision to use an LWIR HgCdTe two-dimensional array coupled to an undeveloped multimodule parallel processor will almost require a higher-cost system than an early architecture decision to use silicon spherical optics, certain processors, and Pt:Si MWIR arrays.

Moreover, engineers doing "clever and creative" designs that satisfy their egos without concern for "producibility, manufacturability, or testability" build significant cost into a product from the beginning. For a low-cost system, remember: *keep it simple, stupid* (KISS).

Source

Rule partially developed by Joe Calabretta, 1995.

LEARNING CURVES

Subject: Management, project management, price/cost estimation

The Rule

Each time a succeeding unit is made, it takes less time to build than did its predecessor. The time can be estimated by a regular decrease every time the number of units made is increased by a factor of two. Mathematically,

$$C_n = C_1 N^{[\,(\log PLR)\,/\,0.3\,]}$$

where C_n = time it takes to make the nth unit in production
     C_1 = time it takes to make the first unit
     N = production number of the unit
     PLR = percent learning rate in decimal notation

Basis for the Rule

This rule was initially based on empirical studies of production lines. It has been applied successfully to all kinds of activities. Usually, manufacturing time decreases by a fixed percentage every time production doubles. Also, most variable costs (e.g., raw materials) follow this rule. The 0.3 in the equation is log 2, because PLR is defined as the improvement experienced when production doubles. If the first unit of a production sensor takes 10 hours to test and the second takes 9 hours, the PLR is 90 percent, and 0.9 should be used in the equation.

Cautions and Useful Range of the Rule

The most significant caution is to be sure that you are applying the correct PLR.
    The above equation assumes that the learning rate is constant; in fact it usually is not. Typically, there is a lower (better) PLR in the beginning as manufacturing procedures are refined, and obvious and easy corrections to the line are applied. As more and more units are made, the rate usually goes higher (less gain). A very mature line may actually experience the opposite effect, wherein it takes longer to make the next unit because of tooling wear and breakage of old capital equipment.
    Some electro-optical programs have experienced a negative learning curve!

Usefulness of the Rule

It is useful in estimating the price, cost, or time needed to build something in production.
    Although the development of learning curves was based on and calibrated by examining touch labor, they can be applied to any variable cost, such as material, and sometimes even to some traditionally fixed costs such as sustaining engineering or management. Therefore, they can be used to estimate total cost and total price directly by using the first unit price in the above equation.

Notes and Explanation

Learning curves are based on the phenomenon that every time you double the numbers of units that you make, you will experience a predictable decrease in the time that it takes you to make each one. This production phenomenon, that each succeeding unit is easier to make, really occurs. The learning curves provide a powerful tool to estimate the reduction in costs or time for a production run. Surprisingly, they can be amazingly accurate. The trick is to know what *PLR* to use. Typically infrared and complicated multispectral systems experience learning curves in the 90 percent range. Visible cameras and simpler systems and components usually have learning curves in the 80s range, and simple mechanical assemblies may even get into the 70s.

Example

Let us state that you develop a product, and it costs $1 million to produce the first prototype after several million dollars of nonrecurring research, development, and engineering (for a production system, these "one-time" costs are to be excluded). Let us also assume that this electro-optical product follows typical learning curves for sensors of ≈ 90 percent and that you are conservative and expect the next unit (the first one for which you charge the customer) to cost as much as the prototype. A quote for five of these is immediately requested. You can assume that it will cost you the following to produce these five:

Production Unit #	Cost per Unit ($)	Total Cost ($)
1	1,000,000	1,000,000
2	900,000	1,900,000
3	846,206	2,746,206
4	810,000	3,556,206
5	782,987	4,339,193

for a total cost of $4.3 million for five units. Because your company exists to make money, you add 20 percent to this figure and respond with a $5.2 million quote.

Now let us say that the president of your company calls you in to inquire about the commercial prospects for your hardware. He wants to know what the thousandth unit would cost to make. You get aggressive with the learning curve for such a large number (assuming the company will invest in automated production facilities) and apply an 80 percent *PLR* and come up with the following projection:

$$C_n = C1 N^{(\log PLR/0.3)}$$

$$C_{1000} = (1,000,000)(1000)^{(\log 0.8/0.3)}$$

and come up with an estimate for the average unit cost of a mere $110,000.

How to Know if Hardware Is in Production

Subject: Management, project management, systems

The Rule

The hardware is truly "in production" if the only problems you have are the name plates and the power supplies.

Basis for the Rule

This rule is based on astute empirical observations.

Cautions and Useful Range of the Rule

Never assume that other things will not go wrong when you are in production— they will!

No matter how many times your production specialists tell you that mundane things (e.g., nameplates) are not an issue, do not believe them.

Usefulness of the Rule

This rule highlights the need to look out for something seemingly minor that can be a critical last-minute holdup to shipping hardware (e.g., missing nameplates) and to keep an eye out for technical problems that pop up in components that typically have producibility issues (e.g., power supplies). Being truly "in production" requires paying attention to the minutiae; do not let your guard down.

Notes and Explanation

Never assume that all the problems have been solved in the preproduction phase of a project. Unfortunately, the preproduction phase usually has only solved performance and other pre-production problems, with little serious attention to manufacturability/producibility problems. Sure, everyone frets over focal planes and optics, but what about those vacuum connectors that have a year lead time, or the space qualified tape, or the special epoxy, or the name plates? Also, unforeseen problems with components frequently occur. These often result from process variations or materials that are not precisely specified. A common occurrence is specifying a minimum number for an attribute without specifying a maximum. In such a case, you can get a transistor with a higher gain (which is a better part), but this causes a change in circuit operation. Others have noted that minor unknown process changes from a supplier, such as insulating a cryocooler case from the ground, or using a different lead coating process of electrical components, can cause strange and inexplicable problems.

Source

Rule provided by Bruce Mitchell, 1995.

THE FEW GENERATE THE MOST RULE

Subject: Management, project management

The Rule

It has long been known that a few individuals generate most of the valuable output.

Basis for the Rule

This rule is based on the historic writings of Keynes, Parkinson, and Augustine on productivity studies, and it can be determined by casual observation.

Cautions and Useful Range of the Rule

It is not always true; sometimes you will have a team of eagles. Unfortunately, sometimes you will have a team of slugs.

Usefulness of the Rule

This rule is useful when doing project management and staffing programs.

Notes and Explanation

Many studies (from sports to the arts and sciences) have shown that most of the useful output of a project is derived from a minority of the participants. In fact, Augustine[1] pointed out that only about one-third of all workers achieve a level of contribution equal to the average of all those who contribute. The trick for the program manager is to find, sign on, and retain these high-performing people. It has been generally noted that increasing the number of participants merely reduces the average output. And, as Ross Perot pontificates, eagles do not flock—you've got to catch them one at a time.

One of the authors (Miller[2]) has noticed a strong corollary to this rule. Most people who are satisfied with their jobs consider themselves to be part of the 10 percent who achieve the most. Most who are not satisfied with their jobs do not consider themselves to be in the top 10 percent. The authors feel that they are both right! Perhaps eagles are all around, but bureaucracy, cronyism, and internal politics are turning most of the workforce eagles into vultures and weasels.

References

1. N. Augustine. 1986. *Augustine's Laws.* New York: Viking, 34–35.
2. J. Miller. 1994. *Principles of Infrared Technology.* New York: Van Nostrand Reinhold, 28–32.

THE EARLY REDUCTIONS ALLOW HIGHER PAYBACKS RULE

Subject: Management, systems, testing

The Rule

Reduction and analysis on data should be done as soon as possible after collecting the data.

Basis for the Rule

This is based on empirical observations, the school of sorrowful experience, and uncommon sense.

Cautions and Useful Range of the Rule

This is a good idea that cannot always be applied due to budget, personnel, or other restrictions.

Usefulness of the Rule

This underscores the importance of timely data reduction and serves as a reminder to add the effort to schedules and budgets.

Notes and Explanation

When data is being taken, it is always good practice to check it for "sanity"; that is, to make sure that it is within an order of magnitude of what was to be expected, and that it is changing in an understood and predictable fashion. Following that, all data (e.g., weather from a space astronomical instrument or a tactical dewar acceptance test) should be completely reduced as soon as possible. Failing to do this will leave one open for potential problems. For instance, much precious time and money can be wasted by collecting garbage data because something simple is wrong with the instrument (cover on) or test setup (wrong wavelength source). It is easier to troubleshoot while you are still at the test set. The operating procedure, instruments, and raw data are immediately available. A week later, they may not be. If you reduce data just before a deadline and find an error, it may be difficult to recover before the deadline.

Source

J. Vincent. 1989. *Fundamentals of Infrared Detector Operation and Testing*. New York: John Wiley & Sons, 241–243.

DOUBLE THE TIME

Subject: Management and estimating schedule

The Rule

Develop a reasonably comfortable schedule for a photonic project and then double it to get a projection of the real time will be required. Alternatively, Augustine[1] points out that projects usually can be completed in a mere 1.3 times the period originally estimated. The same sort of increases are required to estimate the cost to complete the project.

Basis for the Rule

The rule is based on empirical analysis of the grim facts of life.

Cautions and Useful Range of the Rule

This applies to typical E-O programs. The situation probably is not as bad for small, easy projects, and worse for large or complicated projects.

The above rule does not include schedule margins.

Usefulness of the Rule

This is useful for developing schedules, adding necessary margin, and underscoring a basic problem with current project management as applied to EO programs.

Notes and Explanation

Schedules are normally developed on the basis of past experience. However, as entropy dictates, things are falling apart in the universe. Therefore, it will take longer to do a similar task now than it did in the past. Imagine trying to do project Apollo in the 1990s! Unless the task is a simple repeat of a previous effort, it will generally take longer to do than what one anticipates. Occasionally, engineers and managers do develop a reasonable schedule, but rarely does it meet customer or marketing needs, and therefore it is capriciously decreased. The result is the same. Part of the problem is that engineers get bored and always find ways to justify redesigning or modifying a working product. The old rule applies: "If it ain't broke, don't fix it."

Hindsight indicates that this rule is roughly correct for moderate development programs in the tens of millions of dollar range. However, the factor of two tends to underestimate big projects (e.g., the Hubble, Teal Ruby, MSX, advanced NOAA instruments, the space shuttle) where it can reach to 10 to 20.

Augustine points out several examples, dating back to 1798, that seem to support the (apparently) optimistic view that it only takes about one-third more time to complete a project.

Reference

1. N. Augustine. 1986. *Augustine's Laws.* New York: Viking, 158–160.

THE DIVIDE BY THE NUMBER OF VISITS RULE

Subject: Management, project management, subcontractor management

The Rule

Specifications in a data sheet are accurate to the numbers listed, divided by the number of visits that you have had to the supplier.

Basis for the Rule

The rule is founded on empirical observations, the sorry state of the art, and the sorry state of marketing.

Cautions and Useful Range of the Rule

This is another gross approximation by cruel cynics and should be treated as such.

This rule applies to figures of merit that have the property that the higher they are, the better they are (e.g., D^*, optical transmission). The inverse applies when the figure of merit is better when the number is lower (e.g., NEP, cost, weight).

Usefulness of the Rule

This rule is good to remember when estimating performance based on data from a vendor and to underscore a generic serious caution.

Notes and Explanation

Do not believe what you read in marketing sheets. If you are interested, contact the vendor yourself. If you are really interested, buy one of the products and test it yourself (otherwise, you'll be disappointed).

The truth is that data sheets stretch the truth. When you are on the edge of technology in this hurried world, marketing data sheets and catalog descriptions frequently are released before the product is completely designed, or the mass production of it is proven. As a result, frequently the specifications are downright wrong. Additionally, sometimes overzealous marketeers stretch the truth to an extent that would have Gumby screaming in pain. Of course, the specification can usually be met with an increase in cost and schedule.

COST VS. PRODUCTION HISTORY, OR THE MY VIEWGRAPHS WILL ALWAYS OUTPERFORM YOUR HARDWARE RULE

Subject: Project management, aerospace engineering, scientific conferences

The Rule

It is always easier to project low costs or cost effectiveness when no production history exists.

Basis for the Rule

This is unfortunately true and results from overzealous marketeers, empirical observations, and real-world problems in getting state-of-the-art technology to work.

Cautions and Useful Range of the Rule

Since this represents a gross approximation by cynics, care should be always applied. It is usually true, but not always.

Usefulness of the Rule

It is useful to remember this when hearing a pitch for something that still is awaiting a prototype, and the rule underscores a serious caution.

Notes and Explanation

Any competent engineering/scientific crew can make anything work and look appealing. As such, the latest technology from the labs always holds great promise and looks better than anything commercially available (or in the military "field"). Unfortunately, often it is not producible or manufacturable, so it fails horribly when production is attempted (e.g., remember bubble memory?). It is much safer to base a design on something that has a production history or at least a few working prototypes. Obviously, amazing ideas sometimes can be effectively transitioned from the lab to mass production, and the mass-produced item does even better than the lab item (e.g., microelectronics chips). This is called "progress." This is rare.

Although not intuitively obvious, sometimes you can save money by building a prototype (maybe without some of the bells and whistles) before you actually bid on a job. This assumes that the prototype is not a massive project like the Hubble. By investing early in a hardware build, you will probably save more than the cost of the prototype by either deciding to not pursue this project (because the customer cannot afford the real cost, which only you will know) or by properly bidding and convincing the customer that your price is realistic. Additionally, you will have working hardware to impress future customers and develop into other products.

BID THE PRICE TO WIN

Subject: Management, marketing

The Rule

The bid price of an electro-optical system usually can be determined by the following relationship:

$$\frac{P_{c1} + P_{c2} + P_{c3}\ldots [\Delta P]}{S_b} \Rightarrow CH_4 + B_c \times (D_m)^{1/2} = \text{Price}$$

where P_c = your competitors' price
P = your past price
S_b = male bovine waste
CH_4 = methane gas
B_c = crystal ball
D_m = management directive

Basis for the Rule

The rule is based on empirical observations, a cynical view of management theory, and the second law of thermodynamics.

Cautions and Useful Range of the Rule

It works for equally well for systems, component, and study contracts.

Be wary of your perceptions about your competitor's price; they are working as diligently as you to reduce costs.

This applies to price; the actual cost requires quite a different equation.

Usefulness of the Rule

This rule is useful for understanding some of the considerations that may go into your price or the price of your competitors.

Notes and Explanation

This is not all tongue-in-cheek; electro-optics is sufficiently high-tech that the cost is a strong function of "extras." If desired, most systems and components can have costly qualifications, titivations, testing, and features. These may or may not be required for basic operation. Hence, a large part of the price often is based on the (perceived) actions of the competitor, capricious decisions from management, and a tad of bovine waste. Observe that the management directive is only to the 1/2 power. This reflects the fact that, in most technical companies, a management decision can be changed when faced with overwhelming technical evidence.

Source

Rule provided by Grant Milbouer, 1995.

A Cost Index for Space Sensors

Subject: Management, project management, sensor cost, systems

The Rule

The cost of a space based sensors is proportional to the square root of the number of detectors and the aperture area.

$$J_d \text{ is proportional to } (A_o N_d)^{1/2}$$

where J_d = cost index
  A_o = aperture area
  N_d = number of detectors

Basis for the Rule

This rule is based on common sense and empirical observations.

Cautions and Useful Range of the Rule

This provides a crude approximation only because it does not account for the costs associated for spacecraft integration, launch costs, space qualification, and data processing.

Clearly the cost index depends on the maturity of the technology.

Usefulness of the Rule

The above should be used only for a first-cut quick estimate or comparison. It is useful for comparisons of different technologies (e.g., comparing sensors with a HgCdTe FPA versus a silicon camera).

Notes and Explanation

The simplicity of this relationship undermines its worthiness. The cost of any sensor is driven by the aperture size (for large apertures) and the number of pixels. This is especially true of space sensors. The aperture is a good attribute for scaling a sensor's weight and cost. Although lightweight optics reduce the weight effect, they tend to cost more than non-lightweight optics, so it tends to balance out. Likewise, current focal plane arrays produce pixels at a lower cost than the old-style discrete detectors; however, the testing and data processing still scale upward with the number of detectors.

Sources

J. Jamison. 1976. Passive infrared sensors: Limitations on performance. *Applied Optics* 15(4), 891–909.

J. Miller. 1994. *Principles of Infrared Technology.* New York: Van Nostrand Reinhold, 480.

11

Miscellaneous

Well, there is always a place to collect those things that do not fall readily into specific categories. This does not lessen their value, however, and the rules we have assembled here are not exceptions. They include geometric things, such as quick ways to estimate the distance to the horizon and the solid angles of common objects; electronics things, including cooling of electronics and the time it takes to develop circuit boards; imaging things; methods for aggregating the noise in EO systems; and others.

Noise Root Sum of Squares

Subject: Noise, system performance, system engineering

The Rule

Independent noise sources can be added as the root sum of squares. That is, the total variance in the noise of a system is equal to the sum of the variances of the contributing terms.

Basis for the Rule

The random nature of most noise sources encountered in electro-optics allows this to be done. This rule is also derived from a common statistical analysis in which the errors in a system are separately analyzed by taking partial derivatives and eliminating the higher-order terms. Any text on error propagation or analysis of experimental data demonstrates this approach.

Cautions and Useful Range of the Rule

Many other real "noise" sources are not Gaussian, such as microphonics, noise resulting from 60-cycle electronics, or other periodic sources.

This assumes that the sources of the noise are independent.

It does not apply to clutter or other noise sources that may not have a Gaussian property. For example, photon noise from blackbody radiation has a variance equal to the mean photon flux and a standard deviation equal to the square root of the flux. Thus, the noise in a detector will appear as the square root of the sum of the variances $\sigma_1^2 + \sigma_2^2 + \sigma_3^2 + \sigma_{photon}$ where the terms 1, 2, and 3 result from random noise sources.

Care must be taken in developing an error budget in which all terms are assumed to add as the sum of squares. It is quite common for complex optical systems, particularly those with control systems, to accumulate errors in other than a sum-of-squares manner. However, it is almost universally the case that the first analysis is performed using this rule to determine the problem areas. A complete accounting of all error sources and the way they accumulate is a complex endeavor requiring a complete description of the system, which usually does not exist until late in the program.

Usefulness of the Rule

It is useful for calculating noise in an E-O system.

Notes and Explanation

General noise theory indicates that the total noise over a given period of time can be calculated by the root-sum-squared method. That is, one can take the square of all the noise sources and add them together than take the square root of the final sum. On the average, this seems to work very well for typical noise sources encountered in photonic sensors (detector noise, shot noise, dark current, digitizer noise, and so on). However, there are some exceptions, such as clutter, that do not adhere to this rule.

VIBRATION REDUCTION IN SENSORS

Subject: Control loops, servo systems, systems

The Rule

In a sensor that uses a mechanically/optically stabilized image, the control loop should be updated no faster than between 1/5th and 1/10th of the first bending mode of the sensor structure. Stated another way, all vibration frequencies associated with mechanisms in a system should be a factor of ten from the fundamental vibrational frequency of the system structure.

Basis for the Rule

This is a common observation among EO system designers. In practice, they find that the resonant properties of a structure contain many closely spaced modes that result in a system that responds to a wide variety of disturbances. The motion of an image stabilization mirror can stimulate these modes and make the whole system jiggle.

This rule is important in designing all subsystems so that there is little stimulation of the structures of the system. Stimulation of the bending modes results in jitter that has to be removed by electronic or mechanical means, as discussed in another of these rules. It is important to remember that every component in the system has its own structural modes and that this rule should be applied so that the critical ones are not caused to ring through stimulation of their modes. For example, the first bending mode of a large space structure might be as small as a few hertz. Small space structures, such as small satellites, might have a first mode at 10 Hz or above. Optical components can have first modes as low as a few tens of hertz.

Cautions and Useful Range of the Rule

Using this rule will reduce the responsiveness of the system but ensure that the bending modes of the structure are not stimulated. In more advanced and expensive systems, the designer may choose to use a "torque-free" steering mirror system that employs a counter-rotating mass to counteract the forces and torques that would have been induced by mirror motion. Such systems are more complex, and they add mass and cost while reducing reliability.

Usefulness of the Rule

This rule reminds us that a complete modal survey must be completed and complemented with an assessment of the motions that the steering mirror will execute before the real system performance can be estimated. Ideally, an end-to-end simulation of the sensor will be developed, including a combination of optical and structures effects. Unfortunately, there are few modeling environments that allow such assessments without requiring substantial hand work by designers.

Notes and Explanation

Clearly, any disturbance within the sensor that is near a natural frequency of the structure will prevent the sensor from getting full advantage of the stabilization system. That is, the stabilization system will add to the jitter that ultimately degrades the image.

A corollary is that electro-optic sensors cannot support high-bandwidth control systems. If you must control something, use another kind of sensor, such as inertial sensor, which can have bandwidths of several kilohertz.

DEALING WITH POINTING JITTER

Subject: Servo systems

The Rule

Line-of-sight jitter is like dust: it is everywhere all of the time. Generally, jitter that is less than one-tenth of a pixel will not adversely affect the performance of trackers.

Basis for the Rule

Every sensor exhibits these features. Using a binocular or a long focal length lens on a camera easily shows that a human cannot stabilize the line of sight well enough to allow for quality vision. Only if the exposure time of the camera is very short can these systems produce quality pictures.

Usefulness of the Rule

This rule reminds us to constantly consider the effects and control of jitter in an electro-optical system. This requires analyzing the sources of jitter, characterizing the jitter amplitude as a function of frequency, and attention to jitter rejection techniques.

Notes and Explanation

Jitter is the apparent motion of a nonmoving object when viewed by the sensor. Depending on its magnitude and the performance requirements of the systems, jitter may require the installation of hardware that measures and removes it. Some of the removal methods include mechanical stabilization of the line of sight using inertial sensors inside of the sensor, line-of-sight control using the pointing gimbals of the sensor (which is useful only for low-bandwidth correction), and electronic focal plane correction in which each frame is stored and repositioned so that the output image sequence is stable.

The sources of jitter are many. A short list includes

- mechanical platform disturbance
- noisy detectors
- low signal-to-noise ratio
- thermal variations as a function of time

The fact that a perfectly stable sensor will still show jitter results from detector noise. Because the signal that is detected is also contaminated with noise, the output image and its centroid will appear to wander slightly as noise adds to or subtracts from the signal. The signal processing system behind the detector does not know that the variations are due to noise in the detector and interprets them as small angular motions in the target, producing the appearance of target motion when there is none.

SOLID ANGLES

Subject: Geometry, phenomenology, system engineering

The Rules

1. The cone half angle corresponding to unit solid angle (1 steradian) is approximately $(\pi/2) - 1$ radians with an error of 0.2 percent, or about 32.7°.

2. For cones with small angles, the solid angle Ω is approximately $\pi\,\theta^2$, where θ is the half angle of the cone.

3. A 2×2 ft window, 10 ft away, subtends a solid angle of 0.04 sr.

4. The Sun subtends a solid angle of about 7×10^{-5} sr.

5. The ice cream at the top of a cone subtends a solid angle, when viewed from the tip of the cone, of about 0.2 sr.

6. The continental United States subtends about 0.22 sr of the face of the globe. This is obtained directly from the fact that the land area of the U.S.A. is about 9 million square kilometers, and the radius of the earth is 6378 km.

Basis for the Rules

The rules are based on geometric calculations.

Cautions and Useful Range of the Rules

The are to be used only as general guidelines.

Usefulness of the Rules

The rules underscore the size of a steradian and give a feel for solid angles.

Notes and Explanation

Solid angles are two dimensional, as opposed to normal Euclidean plane geometric angles. Solid angles are the two-dimensional projection of a three-dimensional object. Solid angles are generally calculated by dividing the area of the object by the distance to the object squared. Solid angles always enter into radiometric calculations.

Solid angle is defined as $\Omega = 2\pi\,(1 - \cos\theta)$, where θ is the half angle of the cone that forms the solid angle (see Fig. 11.1). Put another way, the solid angle of a cone is equal to the area of the cap of the cone divided by the radius from the vertex squared.

Source

Rules 3, 4, and 5 were adapted from J. Vincent. 1989. *Fundamentals of Infrared Detector Operation and Testing.* New York: John Wiley & Sons, 133.

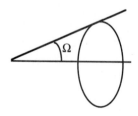

Figure 11.1 Solid Angle

DISTANCE TO HORIZON

Subject: Horizon estimation, phenomenology

The Rule

In statute miles, the distance to the horizon is approximately the square root of your altitude in feet.

Basis for the Rule

The rule is based on Earth-atmosphere geometry.

Cautions and Useful Range of the Rule

It assumes no atmospheric effects, since the rule is based on the geometry of a point positioned just above a sphere.

Usefulness of the Rule

It is useful for quick estimation of the range to the horizon.

Notes and Explanation

The index of refraction depends on the atmospheric density, and the density of the atmosphere depends on the altitude. Therefore, an observer at any altitude can see slightly beyond the geometric horizon.

There are two notes of interest resulting from the differential refraction and dispersion of the atmosphere. The angular extent of the sun is about 0.5°, and the correction of the horizon for refraction can exceed 0.57°. This means that, in some cases, the sun is still visible after it has physically set! The dispersion of the refractive angle causes the refraction correction to be different for different colors. Therefore, the setting sun dips over the horizon one color at a time, with red being first and blue last.

The horizon in kilometers can be estimated by multiplying the square root of your altitude in feet by 1.2. The optical horizon in nautical miles can be estimated by multiplying the square root of your altitude in feet by 1.14.

FRAME DIFFERENCING GAIN

Subject: Image processing, background noise

The Rule

Frame differencing (or temporal processing) can reduce the background and static clutter. Realizable gains are in the realm of a net increase in signal to clutter of about 50 percent for single-order differencing, 80 percent for second-order differencing, and 150 percent for third-order (three frames) differencing. At the same time, the noise per pixel is reduced by approximately the square root of the number of frames processed.

Basis for the Rule

The rule is based on empirical observations.

Cautions and Useful Range of the Rule

Applicability of the rule depends on conditions, background movement, and algorithms.

The performance of frame differencing systems also depends on the stability and jitter of the system.

Usefulness of the Rule

The rule is useful in estimating temporal processing noise gain.

Notes and Explanation

Background subtraction, or differencing one frame from another, can result in almost zero contribution from the static component of background or clutter. However, uncompensated jitter, target trajectory across the background, and changing aspect angles can limit this. The above factors give an empirical/calculated expectations based on the order of differencing.

Source

I. Spiro and M. Schlessinger. 1989. *Infrared Technology Fundamentals.* New York: Marcel Dekker, 282–283.

TIME REQUIRED TO DESIGN A CIRCUIT BOARD

Subject: Electronics, systems

The Rule

Usually, five to nine designer-months are required for engineering and documentation for each electronics board associated with an electro-optical system.

Basis for the Rule

The rule is based on empirical observations that have been made in the context of the electronics that are associated with electro-optical systems.

One might expect that this time interval would decrease with the advent of advanced chip technologies. The problem is that, as more capability is made available in each chip, designers want to make their products more capable. That results in the inclusion of microprocessors, advanced input-output devices, and other complex technologies. In addition, few systems are built without having to include ROM-based software that adds substantially to the time required to develop the product. The result is that board development times remain the same. At the same time, their costs are going up.

Cautions and Useful Range of the Rule

This is an average; simple boards require less time, more complicated boards (especially mixed analog and digital) require more time.

If the design is largely done or breadboards exist for another program, then you are lucky and can complete such a board faster.

This assumes that you can obtain a reasonable response time from board makers, chip manufactures, and ASIC developers. Increasingly, these foundries are ignoring those who make scientific, military, or other small-volume electronics, concentrating instead on the glitz and profit of commercial products.

It is possible to make a baby in less than nine months. To do so, start with a pregnant woman. Likewise, it is possible to get an electronic board quickly. Start with one that is partially developed, or modify an existing one.

Usefulness of the Rule

It is useful for estimation the effort and time needed to build a board.

Notes and Explanation

Generally, an electronic board is composed of components mounted on a printed circuit board. Recently, the SEM-E and VME circuit board architectures have been the most acceptable. Even using these standard formats, it seems that between five and nine designer-months are necessary to produce a working board from scratch.

ELECTRONIC BOARD COOLING

Subject: Electronics, systems, cooling

The Rule

The cooling method required for an electronics board depends on the number of watts dissipated, its ambient operating environment, and its size. Generally, for standard board sizes,

- < 25 W   no cooling necessary if operated in a standard atmosphere
- 25–50 W   forced-air cooling required
- 50–100 W   liquid cooling required
- >100 W   generally too much waste heat, junctions are likely to melt

Basis for the Rule

These assumptions are based on empirical observations along with a generalization of thermal engineering.

Cautions and Useful Range of the Rule

The rule assumes continuous operation and power dissipation from the board, operation at typical military ambient temperatures (55° C or less), atmospheric pressure of at least 400 Torr, and board dimensions of ≈ 13 × 23 cm. It assumes standard board construction and formats (e.g., VME).

Usefulness of the Rule

This is useful for a quick estimation of the number of components that can be placed on a board, given the level of cooling available.

It is also useful for estimating the type of cooling needed for a board.

Notes and Explanation

When designing or implementing a new electronics board, the above power levels should not be exceeded for the given cooling method. Otherwise, reliability will decrease rapidly, and junction temperatures may get hot enough to melt the solder. These numbers would be lower for space applications where there is no natural convection.

CHIP YIELD

Subject: Electronics, systems, cost, yield, FPAs

The Rule

Chip yield is inversely proportional to the size of the chip and defect density raised to the number of masking steps, or the yield of an photolithographic process is proportional to

$$\text{Yield} \propto \frac{1}{(1 + DA)^{m}}$$

where Yield = expected yield of a die from a wafer
    D = expected defect density per square centimeter per masking step (If you do not know, assume one defect per square centimeter for silicon, and more for other materials.)
    A = area of the chip in square centimeters
    m = number of masking steps

Moreover, Murphy[1] developed the following model for the probability that a given die from a wafer will be good:

$$P_{g} = \left(\frac{1 - e^{-AD}}{AD} \right)$$

where P_{g} = probability that a die is good after the processing of a wafer

Additionally, Seeds[2] gives us the following equation:

$$P_{g} = e^{-\sqrt{AD}}$$

Basis for the Rule

The rule is based on industrial engineering and an empirical analysis of the grim facts of life applied to yield phenomena.

The above equations are based on processing silicon; other materials (e.g., GaAs, InSb) will have lower absolute yields but should approximately follow the same curve.

The latter two equations assume a Poisson distribution of defects in the wafer.

Cautions and Useful Range of the Rule

Calculating yield is a complicated process; foundries typically have complex models for this. Yield is also sensitive to design layout, design methodology, type of circuit, production line quality, and the age of the manufacturing equipment. However, the above equations provide a good approximation with the assumptions that the process is well controlled and developed, assuming that crystal defects are low and breakage and human handling do not contribute.

One must use the right numbers for defect density. This is not always trivial, as it can depend on not only the obvious (rule size, material, material purity) but subtle process attributes (individual machine, operators, the kind of night the operator had before, and even room temperature and humidity). All of these models tend to give small probabilities of success when typical silicon defect densities are used (1 to 2 per cm^2). It is probably best to scale these with a constant based on experience in producing a similar IC, which is indicated by the "$\propto$" in the first equation.

Technology is always improving, so the form of these equations may change with the turn of the century as processes improve, as implied by the lack of the masking steps in the second two equations.

Some defects can cause a circuit to be inoperable but do not affect the interconnects. Generally, yield is better when considering the fact that some of the chip is made up just of interconnects, and by ignoring the edges of the wafer (where chips are not made and crystal defects are larger).

Often, the exotic materials used by EO engineers have wafer sizes much smaller than silicon (e.g., 4 cm diameter for HgCdTe as opposed to 10 cm for silicon [a factor of over 6 in area]); therefore, the wafer size and effects on final product yield should be considered for these materials.

Usefulness of the Rule

These equations are useful to the designer to underscore the drivers on yield.

In addition, they are valuable in estimating the yield, cost, schedule, and agony of a new chip design. Cost can be directly applied, as it varies almost in direct proportion to the wafer processing yield. Generally, one can assume a processing cost of about \$1.00 to \$2.00/cm^2 for silicon (a factor of $\approx$ 10 to 10,000 higher for detector materials). Many experts believe that the cost per square centimeter for Si is not likely to decrease much in the upcoming years. This is because future cost reduction in Si chips is likely to occur from decreased feature size, added functionality per chip, and increased wafer size rather than reduced wafer processing costs.

Notes and Explanation

In summary, Table 11.1 delineates the impact of chip design attributes on yield.

References

1. B. Murphy. 1964. Cost size optima of monolithic integrated circuits. *Proc. IEEE, vol.* 52, 1537–1545.
2. R. Seeds. 1967. Yield and cost analysis of bipolar LSD. *Proc. IEEE International Device Meeting,* 12.

Additional Sources

R. Geiger, P. Allen, and N. Strader. 1990. *VLSI, Design Techniques for Analog and Digital Circuits.* New York: McGraw-Hill, 19–27.
J. Miller. 1994. *Principles of Infrared Technology.* New York: Van Nostrand Reinhold, 124–125.

TABLE 11.1 Chip Design Attributes vs. Yield

Attribute	Impact on Yield	Explanation
Die (or chip) size	Increasing size greatly reduces yield.	As the die size increases, it is more likely that a defect will occur.
Ruling size	Decreasing ruling size (lower than ≈0.8 μm) increases yield.	The smaller the features, the more critical the alignment and statistical variations in process. Historically, yield has greatly improved with smaller rule size. Eventually, small features (approaching the de Broglie wavelength) will produce higher yield. However, one can expect that high yields from the 0.1-μm process will not be achieved before 2005.
Mask steps	The more masking steps, the lower the yield.	Alignment errors and dust on masks reduce yield.
Analog vs. digital vs. hybrid	Digital circuits have the best yield, analog the next best, and hybrid the worst.	Statistical process variations cause the miniature circuits of an IC to have variations. For a digital circuit, these often do not cause the output to change (e.g., a "0" to flip to a "1") while they add directly to analog errors.
Material	Material greatly affects yield.	Silicon is the most mature material, with the highest yield. GaAs is the next most mature, and everything else implies low yields.
Circuit size	Smaller circuit areas reduce yield.	Smaller circuit size requires smaller rulings (see above) and requires better alignment of masks and increased process control.
Defects/area	This factor greatly affects yield.	The manufacturing process, from crystal growth to chip packaging, is affected by the number of defects/area.

NOISE DUE TO QUANTIZATION ERROR

Subject: Electronics, control systems, digitizers

The Rule

The effective noise due to a quantization error is:

$$Q_n = \frac{LSB}{\sqrt{12}}$$

where Q_n = quantization noise (e.g., in electrons)
LSB = value of the least significant bit (in electrons)

Basis for the Rule

The rule is based on statistical data reduction.

Quantification error is uniformly distributed. The standard deviation of a uniform distribution is its amplitude divided by $\sqrt{12}$.

Cautions and Useful Range of the Rule

The LSB value is small compared to the total noise. The rule is an approximation only.

Usefulness of the Rule

This is useful for noise estimation, setting dynamic range, and determining the number of bits of A/D needed.

Notes and Explanation

Typically, the original analog output from a detector is transformed into a digital stream. This digitization has a finite resolution based on the desired total dynamic range and the number of bits of the A/D employed. For some system designs and circumstances, this can be a serious contributor to noise.

If the least significant bit is small compared to the total noise, then the probability that it will fall between $-LSB/2$ and $+LSB/2$ is roughly constant, and the RMS error reduces to the above relationship. This noise becomes more unpredictable as LSB approaches the basic noise of a system.

Chapter

12

Ocean Optics

The interaction of light and water has been a topic of study for many centuries, and for good reason. It is well understood that light in the ocean stimulates the microscopic plant life that supports the food chain and ultimately defines the availability of food resources for man. The earliest interest related to the characteristics of vision when the observer is submerged or is viewing submerged objects. These problems were successfully managed when the theory of refraction was understood. By the 1940s, Duntley had begun his pioneering work on the optical properties of clear lake waters. Preisendorfer assembled the existing theory in the mid 1970s and thoroughly summarized the state of the theoretical nature of the problem.

The introduction of the laser also stimulated additional work on light propagation in the ocean. Hickman pioneered the use of pulsed lasers to measure water depth in coastal regions and to define the environmental properties of ocean and coastal waters. One of the authors of this book (Friedman) spent a number of years working with Hickman and his team in the characterization of surface pollution, using fluorescence and other techniques for detection of oil, algae, and environmental contaminants. Much of that work relied on traditional characterizations of the aquatic environment, including the absorption, scattering, and total attenuation coefficients. The researchers also employed a variety of instruments for characterizing water in both natural and laboratory environments. One of the important instruments in the arsenal was the absorption meter, developed by contractors and employees of NASA and the U.S. Navy. In addition, the advent of the environmental movement led to the realization that optical properties of water could be a sensitive indicator of its quality, the nature of the suspended sediments in it, and the presence of biological and hazardous materials. As a result, agencies such as the United States Geological Survey and others developed optical methods for water quality definition.

The development of the laser also added greatly to interest in and the success of underwater instrumentation. The recognition that beam attenuation coefficient would be as important as diffuse attenuation coefficient in ocean optics caused

NASA, among other government agencies, to begin research programs to improve the interpretation of remote sensing data.

The high brightness of lasers and the ability to select wavelengths that propagate well in the ocean led to attempts to develop imaging and target tracking systems. This required developments in both the theory of radiation propagation in turbid media and the creation of new optical systems able to cope with the low temperatures and high pressures associated with operating deep in the ocean. The result has been a great leap forward in underwater imaging sciences of all types. Television cameras are regularly used at thousands of feet of depth.

In a related development, the properties of the ocean were evaluated to determine the impact of the transparency of the ocean on remotely sensed images taken from space platforms. Knowledge of ocean optics has become essential in the proper interpretation of images that include water scenes.

The U.S. Navy has a long-standing interest with the use of lasers for communication with submarines. This has stimulated considerable work on the properties of the ocean, as well as the development of special lasers and receivers that match wavelengths well within the optical window of water. In addition, there has been considerable effort to understand how to take advantage of the spectral purity of lasers to perform underwater imaging that is not possible from conventional light sources. In all of these cases, both experimental and theoretical work has been performed to determine the limits of performance imposed by the natural water turbidity and the sensitivity of the system to the ratio of absorption to scattering in the medium. The latter effect has a potentially degrading impact on imaging capability because light emanating from the object does not take a direct path to the viewer. The absorption part of the problem has a greater impact on the amount of radiation that can reach a particular distance but does not affect imaging.

The reader interested in finding new information about this field should concentrate on reading the various SPIE compendiums of papers that focus on ocean optics. They have had a common title for years: *Ocean Optics*. One recent edition covers a convention on the subject that was held in Norway in June of 1994. Most of these papers are presented at a fairly sophisticated level, requiring that the reader have some familiarity with the field. Occasionally, one finds an oceanography book that provides a good foundation. An example is the work by Apel. This is a field of EO that gets relatively little attention, so the interested reader will have to do some digging to find new ideas and instrument descriptions.

INDEX OF REFRACTION OF SEA WATER

Subject: Ocean optics, index of refraction

The Rule

The index of refraction of sea water depends on its temperature and its salinity. It can be approximated by

$$n = 1.33 + [3400 + n_1 T + n_2 T^2 + n_3 T^3 + S(n_4 + n_5 T + n_6 T^2 + n_7 T^3)] \times 10^{-6}$$

for a wavelength of 589.3 nm where the following coefficients are used:

$n_1 = -0.86667$
$n_2 = -0.2350$
$n_3 = 1.16667$
$n_4 = 19.65$
$n_5 = -0.1$
$n_6 = 2.25 \times 10^{-3}$
$n_7 = -2.5 \times 10^{-5}$

T and S are the temperature and salinity, respectively.

Basis for the Rule

This is an approximate polynomial expansion based on a curve fit to empirical data.

Cautions and Useful Range of the Rule

This approximation compares favorably with the real index of refraction of water, as shown by Collins.[1] It also assumes that the wavelength is 589.3 nm.

Usefulness of the Rule

This rule provides an easy method for estimating the change in index as a function of temperature and salinity. The index of refraction is a key factor in determining the reflectivity of the ocean.

Notes and Explanation

The index of refraction of water is a significant factor in interpreting vision and imaging in the ocean and is important in properly interpreting remotely sensed data.

Reference

1. D. Collins. 1984. Recent progress in the measurement of temperature and salinity by optical scattering. *Ocean Optics VII,* SPIE, vol. 489. Bellingham, WA:, SPIE.

f-STOP UNDER WATER

Subject: Ocean optics, underwater photography

The Rule

A good rule of thumb for underwater photography is that the f-stop should be changed one stop (the effective aperture increased) for every 36 ft of increased distance from the subject.

Basis for the Rule

This rule has been observed in real ocean conditions. It could be guessed by the following argument. The beam attenuation coefficient in very clear water, the type in which photography will be attempted, is about $0.05\ m^{-1}$. This means that any light emanating from the object to be photographed will be reduced in intensity a factor of two in about 12 m, which is about 36 ft.

Cautions and Useful Range of the Rule

Photography in any conditions, including in the ocean, can be affected by scattered light, unnoticed sources of light, conditions of the target, and so on. Therefore, it is best in all cases to determine the most likely exposure and then bracket it with exposures one f-stop above and below the most likely value. The same applies here. In fact, since most underwater photographic opportunities are rare, it would be appropriate to shoot a series of exposures that vary from 2 stops below to 2 stops above the most likely value.

This rule applies to very clear water and underestimates the amount of light that will be available.

Usefulness of the Rule

As stated above, this rule applies for photography in clear water. This is almost always the condition under which photos will be obtained, given that cloudy and turbid water will not only attenuate the propagation of light but will have very low contrast due to extensive scattering in the medium.

Notes and Explanation

Water is an exponential medium from the point of view of light transmission. This means that both directed beams of light, as well as diffuse fields of light, are attenuated exponentially with distance. As a result, the normal methods of estimating the f-stop required for proper exposures in photography do not apply as in the case of air.

Source

L. Mertens. 1970. *In-Water Photography.* New York: Wiley Interscience, 29.

ABSORPTION COEFFICIENT

Subject: Ocean optics, absorption

The Rule

The optical properties of water include the scattering and absorption coefficients, which, when summed, give the total attenuation coefficient for collimated light. The absorption coefficient, a, can be estimated from the total diffuse attenuation coefficient, K, by

$$a \approx \frac{3K}{4}$$

Basis for the Rule

Empirical evidence, based on the propagation of beams and plane waves, such as sunlight, have resulted in a series of approximations that let one ocean optics parameter be derived from other measurements. The absorption coefficient, a, is a measure of the energy loss in a beam propagating in the ocean due to absorption by water and its suspended constituents. The absorption coefficient is used in Beer's law to compute the intensity of the beam after traveling a distance in the water.

Cautions and Useful Range of the Rule

Water types and optical conditions in the ocean vary significantly worldwide. However, this rule can be useful in setting up the dynamic range of beam attenuation instruments based on diffuse attenuation measurements.

Usefulness of the Rule

It is frequently hard to obtain the absorption coefficient for a body of water, since to do so requires a special instrument such as the one developed by Friedman et al.[1] Instead, when some inaccuracy is allowed, this rule can be used.

Notes and Explanation

$I_L = I_o \exp(-a\,L)$ when there is no scattering present. In the equation, I_o is the intensity of the beam at the starting point and I_L is the intensity at distance L. K is measured using a wide-field instrument that is suspended in the ocean and collects light from the hemisphere above it. By measuring the intensity as a function of depth, the value of K can be determined because it, too, is the scale factor in Beer's law for diffuse light. That is, when K is measured, scattered light that remains in the downwelling field is collected, unlike the case when beam measurements are made.

Reference

1. E. Friedman, L. Poole, A Cherdak, and W. Houghton. 1980. Absorption coefficient instrument for turbid natural waters. *Applied Optics* 19(10).

Additional Source

H.R. Gordon et al. Introduction to ocean optics. 1984. *Ocean Optics VII*, SPIE, vol. 489. Bellingham, WA: SPIE, 36.

ABSORPTION DUE TO CHLOROPHYLL

Subject: Ocean optics, absorption

The Rule

The following rule can be used to estimate the additional absorption due to chlorophyll, over and above that contributed by the water itself:

$$\text{Absorption due to chlorophyll} = 0.0667(C_{g/L})^{0.758} \ m^{-1}$$

where $C_{g/L}$ = chlorophyll concentration in micrograms per liter

Basis for the Rule

This rule results from measurements in the North Atlantic Ocean. Data were assembled from measurements at a variety of depths in the euphotic zone.

Cautions and Useful Range of the Rule

The wavelength to which the equation applies is 670 nm. Yentsch and Phinney[1] have described the method for measuring the concentration of plankton absorption by using a filter to collect the plankton samples. They note that other techniques have been suggested.

Usefulness of the Rule

This rule, and others like it, can be effective in estimating the penetration of light into the ocean, which can be used with remote sensing information to estimate primary production of small aquatic plants and unicellular algae.

Notes and Explanation

It is well known that phytoplankton in the ocean contributes to optical absorption. The methods employed to measure the absorption are subject to some debate but, in the final analysis, the methods employed by Yentsch and Phinney, and the results they have obtained, will be useful to those who use remote sensing data to estimate concentrations of algal "blooms." Some of the methods involve reflectance measurements of algal samples, whereas others use mechanical or chemical methods for extracting the pigment-bearing part of the cells and doing spectroscopy on the resultant solution.

It should be pointed out that the data they have obtained include mixes of algal types, and that blooms tend to contain high concentrations of a single specie, thus meaning that the exact values of the coefficients presented in the rule might need to be modified on a specie-by-specie basis.

Reference

1. C. S. Yentsch and D. A. Phinney. 1988. Relationship between cross-sectional absorption and chlorophyll content in natural populations of marine phytoplankton. *Ocean Optics IX*, SPIE, vol. 925. Bellingham, WA: SPIE, 109.

OCEAN REFLECTANCE

Subject: Ocean optics

The Rule

Given the diffuse backscatter coefficient, b_b(meters^{-1}) and the absorption coefficient, a(meters^{-1}), we can estimate that the irradiance reflectance, R, is[1]

$$0.33 \frac{b_b}{a}$$

Basis for the Rule

A more complete computation gives

$$\frac{\dfrac{b_b}{a}}{1 + \dfrac{b_b}{a} + \sqrt{1 + 2\dfrac{b_b}{a}}}$$

The result of the surface reflectance is an upwelling radiance, just above the surface, of[2]

$$\text{Upwelling radiance} = \rho\, L_{sky} + \frac{t}{n^2} L_u$$

where ρ = surface specular reflectance coefficient
$\quad L_{sky}$ = radiance (W/m^2) on the ocean surface due to sky light
$\quad\quad t$ = transmittance of the air-ocean interface
$\quad\quad n$ = index of refraction of the water
$\quad L_u$ = upwelling light field in the water (W/m^2)

Cautions and Useful Range of the Rule

A knowledge of some characteristics of the scattering of suspended materials can be used to estimate gross features of ocean optics.

These rules assume that the optical properties, absorption and scattering, are linear functions of the concentration of plankton. This assumption may be stretched to the limit as one considers the spectral properties of the various plankton types. In addition, the shorter version of the rule is easier to use but is not as accurate as the more complex form.

Usefulness of the Rule

This rule provides an easy an quick estimate of the upwelling light field just above the surface. This allows the designer of ocean optical instruments to define the likely requirement for dynamic range and provides an estimate of the change in the upwelling light as a function of ocean conditions. This allows the designer to improve estimates of the light intensity beyond that which results from simply considering the ocean reflectance.

Notes and Explanation

Remote sensing of the ocean is a valuable economic and scientific technology. Proper interpretation of the results involves removing the effects of the intervening atmosphere, correcting for the effect of waves and clouds, and proper interpretation of the results of those corrections. A key part of the interpretation is the recognition of the presence of phytoplankton, which impose a spectral content on the signatures obtained in the imagery. The importance of these corrections and interpretations is a strong function of the spectral resolution of the images. For example, many of the detailed effects of plankton on the image spectral content will be lost in LANDSAT or SPOT multispectral data. However, hyperspectral imagery, in which up to hundreds of spectral bands may be recorded, will show the presence of the plankton, and more advanced interpretive methods will be needed.

References

1. J. C. Erdmann and J. M. Saint Clair. 1988. Simulation of radiometric ocean images recorded from high-altitude platforms. *Ocean Optics IX,* SPIE, vol. 925, Bellingham, WA: SPIE, 36.
2. H.R. Gordon et al. 1984. Introduction to ocean optics. *Ocean Optics VII,* SPIE, vol. 489, Bellingham, WA: SPIE, 40.

UNDERWATER DETECTION

Subject: Ocean optics

The Rule

For most observers with experience in attempting to find submerged objects, it is found that the object can be observed at a distance computed from the following expression:

$$\frac{5}{\alpha - K\cos\theta}$$

where α = attenuation coefficient for collimated light (meters^{-1})

K = diffuse attenuation coefficient (meters^{-1})

θ = zenith angle measured from the swimmer [The notation used by Preisendorfer[1] is a "swimmer centered direction convention;" that is, a downward view corresponds to a zenith angle of 180°. A horizontal view has a zenith angle of 90°, so the scaling range is $1/\alpha$.]

Basis for the Rule

This is the result of empirical investigations and will vary somewhat with the abilities of the observer. However, Preisendorfer's work on this subject is exhaustive and provides a wide range of examples related to visibility and biology in the marine environment. Many of the concepts are derived from the theory of radiation transport in turbid media and are beyond the scope of this book.

Cautions and Useful Range of the Rule

A quick review of Preisendorfer's book shows that he covers a wide variety of viewing and lighting conditions. This rule applies in general and can be used as a first approximation. As is always the case in analysis of human vision, the exact conditions of any particular situation must be analyzed in detail.

Usefulness of the Rule

This rule provides a first guess when considering the ability of a submerged swimmer practiced in sighting objects to detect objects underwater.

Notes and Explanation

Underwater vision is of aesthetic and practical interest. Early interest in underwater optics did not have the advantage of the advancements in EO technology that have occurred in the last 40 years or so. This area of geophysics now has become a fairly mature science with a full range of theoretical and experimental results.

Example

In extremely and exceptionally clear water near the surface, with an α of 0.1 m^{-1}, a swimmer looking horizontally can see about 50 m (5/0.1). Note that, in this case, we assume the swimmer is looking horizontally, thus making $\cos\theta$ equal to zero.

Reference

1. R. W. Preisendorfer. 1976. *Hydrologic Optics*, vol. 1. Washington, DC: U.S. Dept. of Commerce, 194.

WAVE SLOPE

Subject: Ocean optics, wave slopes

The Rule

From Fraedrich,[1] the mean square surface slope is defined by its variance (σ^2) and can be approximated by

$$\sigma^2 = 0.003 + 5.12 \times 10^{-3} W$$

where W = wind speed

Basis for the Rule

The equation is empirical.

Cautions and Useful Range of the Rule

Apel[2] shows that surface wind speeds over the globe range up to about 10 m/s, except in storms, where it can be much higher.

   The equations that define the slopes are given below.

Usefulness of the Rule

The apparent surface reflectance of the ocean depends on the combined effects of material reflectance and the range of slopes of the surface. Ocean reflectance is a critical factor in the performance of a number of remote sensing systems. Knowing the effective reflectance allows the system designer to better estimate the contribution that ocean reflectance will make to the radiation reaching the sensor. This rule allows the influence of the wind to be factored into the analysis.

Notes and Explanation

A truly flat surface exhibits a mix of specular and diffuse reflectance, with the latter resulting from the subsurface scattering that occurs. In the presence of wind, the surface takes on a new character and exhibits glint.

   Fraedrich also gives the following equations:

$$\sigma^2 = (\ln W + 1.2) \times 10^{-2} \text{ for } W < 7 \text{ m/s}$$

$$\sigma^2 = (0.85 \ln W - 1.45) \times 10^{-1} \text{ for } W > 7 \text{ m/s}$$

Using these approximations, the time-averaged radiance of the ocean can be estimated using methods defined in Ref. 2 but too complex to be included here.

Reference

1. D. Fraedrich. 1988. Spatial and temporal infrared radiance distributions of solar sea glint. *Ocean Optics IX*, SPIE, vol. 925. Bellingham, WA: SPIE, 392.
2. J. R. Apel. 1987. *Principles of Ocean Optics*, Orlando, FL: Academic Press, 201.

UNDERWATER GLOW

Subject: Ocean optics

The Rules

A persistent source of light in the ocean is due to the action of biota. Resulting from luminous plankton, bioluminescence has the following properties:

1. At night, the bioluminescence follows the temperature, with the maximum lighting occurring in the mixed layer and with decreasing intensity below the thermocline.

2. There is significant diurnal variation in the light intensity.

3. The spectrum in surface waters ranges from 360 to 620 nm, with a peak around 480 nm.

Basis for the Rules

This set of rules is the result of empirical observations made on cruises in the Pacific Ocean.

Cautions and Useful Range of the Rules

These rules are general enough that they will almost always be correct. The data have been developed in a series of measurements in the Pacific, Atlantic, Barents, and Mediterranean seas. Data have been obtained from a variety of depths, including the near-surface region (around 200 m depth) to depths in excess of 3600 m. In the latter case, the presence of a deep-diving submersible is expected to stimulate the light emission from the various organisms in the sea. Clearly, one can imagine that the actual light emission might vary, depending on the velocity, size, and turbulence generated by an object deep in the ocean. Therefore, the specific spectra and radiance levels that were observed may be very much the result of the details of how the experiment was performed.

Usefulness of the Rules

These types of general rules have the purpose of keeping the EO designer aware of the presence of biological sources of light. Observations made in the natural environment will have such lights as a part of the background that sensors will encounter.

Notes and Explanation

The field of ocean optics has a number of applications in remote sensing for commercial and military purposes. The types of data presented in this rule tend to deal with the case in which observations are being made in the ocean, since the excitement of bioluminescence presumes that something like a submarine or other object is moving through the water. Designers of camera systems or other types of imaging methods need to take this additional source of radiation into account when computing background levels that might be encountered. That is, one can estimate the amount of sunlight present as a function of depth using the diffuse attenuation coefficient, K. The biological sources of radiation must be considered as well, because their presence will reduce the contrast observed in imaging of submerged objects, using either residual sunlight or artificial light sources.

Source

J. Losee et al. 1984. Bioluminescence in the marine environment. *Ocean Optics VII,* SPIE, vol. 489. Bellingham, WA: SPIE, 77.

13

Optics

Optics tends to be a discipline whose state of the art is advanced by the needs of users. Generally, development in optics seems to have been linked to specific engineering applications. Optics of antiquity, until about 1700, existed mainly an aid to vision. From about 1600 to the World War II, the main impetus for optical development was to develop better instruments for astronomy. Navigation and vision aid played a key role in this era, but most major developments were somehow geared to astronomy (e.g., the Foucault knife edge test, interferometers, new telescopes, and so on). Military needs dominated optics development from World War II to the 1990s. In the 1990s, the military faded as the driving force, to be replaced by communication and computing. It can be estimated that in a few decades (say, 2020) the emphasis will again shift. It is conjecture to predict what the driving force will be, but it might be something like bionics or robotics.

The science of optics began thousands of years ago. Archeological findings involving the Phoenicians suggest that powered lenses were made over 3000 years ago. Clearly, all of the ancient cultures studied light and its interaction with matter. Aristotle, Plato, Ptolemy, Euclid, Pythagoras, and Democritus all wrote extensively about vision and optics. Seneca (4 B.C. to A.D. 45) was the first to write about observing light divided into colors by a prism. To these early investigators, the world was full of rules-of-thumb and principles explained by thought process alone. Sometimes this resulted in poorly made or poorly understood observations. Among the incorrect theories was that vision resulted from "ocular beams" emitted from the eye. This theory was finally rejected by al-Kindi and al-Haitham.

The first painting of a person wearing glasses is attributed to Tommaso of Medina's painting of Hugues de St. Cher in 1352; however, spectacles and lenses were known to glassmakers for several centuries prior. No one knows who invented spectacles. Likewise, much controversy surrounds the inventor of the first telescope, although it probably occurred around 1600 by Lippershey (who applied for a patent in 1608), Adriaanzoon, Jansen, or someone else. Telescopes were being sold as toys and navigation aids while Galileo and others turned them to the heavens for astronomy. Prophetically, Galileo remarked that the science of astronomy would improve with further observations from better telescopes. The microscope was invented about the same time, with almost as much controversy.

Theory followed inventions and the new observations that they provided. A few decades later, in the mid-1600s Snell, Descartes, and Huygens were working on the law of refraction, which became known as Snell's law (or Descartes law). The publication of Isaac Newton's *Opticks* in 1704 was a milestone in the science and engineering of optics and started the argument of whether light was a stream of particles (corpuscles) or a wave. It behaves as both. A century later, Thomas Young developed his double-slit experiment indicating light was a "undulation of an elastic medium." Conversely, a little more that another century later, Einstein won his Nobel prize for analyzing the photoelectric effect.

The seventeenth century also saw the development of the reflective telescope promoted by Marin Mersenne. In the mid-1600s, James Gregory developed the Gregorian configuration, Guillaume Cassegrain the Cassegrain, and it is generally accepted that Isaac Newton built the first useful reflective telescope in 1668. Incidentally, when referring to the Cassegrain design, Newton provided a glimpse into his jealousy by boldly stating that "the advantages of this device are none."

Then, in 1800, William Hershel (1738–1822) reported the discovery of light beyond the visual spectrum and attempted to determine the radiant power as a function of wavelength. Shortly afterward, Fresnel and Fraunhofer developed the diffraction theory, which added greatly to the discipline of optics and provided the basis for many of the following rules. The late 1800s were marked by Maxwell and Michelson putting to rest the theory of the ether.

This century saw the development of manufacturing technology, including John Strong's advancements in reflective coatings, the invention of the laser, and development of holography. The latter half of the 1900s was the age of the application of optical sciences to other fields such as electronics (photolithography making the integrated circuit possible), medicine, nondestructive testing, spectroscopy (for chemical analysis), and a great deal more.

The reader who is interested in more detail about optics has a surplus of available texts. The authors would suggest Hecht's *Optics* as a first review. Following that, for specific EO applications, the following texts provide excellent (although somewhat parochial) discussions in their associated chapters: Hudson's *Infrared engineering,* Spiro and Schlessinger's *Infrared Technology Fundamentals,* and Miller's *Principles of Infrared Technology.* For more detailed analytic discussions, one should seek Fowles' *Introduction to Modern Optics* and the venerable Born and Wolfe's *Principles of Optics.* Additionally, there are many series and handbooks available on this diverse subject. For an understandable discussion of the quantum electrodynamic fundamentals of optics (e.g., to learn what really occurs in reflection, refraction, and the like) one should review Feyman's *QED.* The academic journals that typically specialize in this discipline include *Applied Optics,* the *Journal of the Optical Society of America,* and *Optical Engineering.* For late-breaking news of the technology, one should consult *Laser Focus, Photonics Spectra, Physics Today,* and *Sky & Telescope.* Several professional organizations frequently hold seminars (and publish their associated proceedings), including SPIE, IRIA's IRIS, and AIP/OSA. There are several optics "users groups" on the Internet. Also, do not forget the optics catalogs and users guides available from manufacturers. Several have excellent engineering discussions of pertinent principles and available technologies.

THE LAW OF REFLECTANCE

Subject: Optics, optical design

The Rule

When light encounters a smooth, specular reflective surface, it is reflected at an angle equal to its incident angle with the normal to the surface (see Fig. 13.1). The incident, reflected, and surface normal are in the same plane.

Basis for the Rule

This rule is based on Fermat's principle, simple geometric optics, and empirical observations.

Cautions and Useful Range of the Rule

This is valid throughout the electromagnetic spectrum.

It assumes that the surface is large compared to the wavelength of light and is valid for highly polished, reflective surfaces that are smooth compared to the wavelength (specular).

Usefulness of the Rule

This forms the basic for all reflective ray tracing and reflective optical design.

Notes and Explanation

When a light bundle encounters a reflective surface, most of the light that is reflected is done so at the same angle from the normal as is the incident ray, but directly opposite to it.

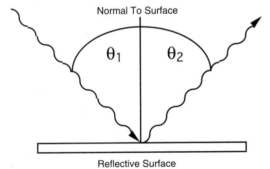

Figure 13.1 Reflection from a Specular Reflective Surface

LAW OF REFRACTION, OR SNELL'S LAW

Subject: Optics, lenses, and windows

The Rule

When light encounters a surface of a different index of refraction, it is refracted (see Fig. 13.2) in a relationship of:

$$n_1\sin\theta_1 = n_2\sin\theta_2$$

where n_1 = index of refraction of the first media
θ_1 = angle of incidence from the first media
n_2 = index of refraction of the second media
θ_2 = resulting angle in the second media

Basis for the Rule

This rule is based on Fermat's principle, basic geometrical optics, and empirical observations originally included by Snell. It can also be derived from the Principle of Least Action as illustrated in a number of basic physics texts.

Cautions and Useful Range of the Rule

Snell's law is valid throughout the electromagnetic spectrum. It assumes that all of the angles considered above lie in the same plane.

Usefulness of the Rule

This rule forms the basics for all refractive ray tracing and refractive optical design. It is valuable to be aware of it when considering any refractive element.

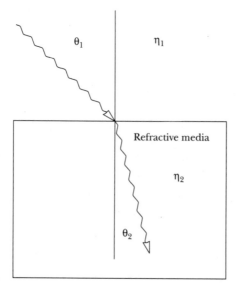

Figure 13.2 Reflection from a Specular Reflective Surface

Notes and Explanation

W. Snell and R. Descartes both stated this rule in the first part of the 1600s. This law also determines when "total internal reflection" occurs (i.e., when one part of the above equation equals or exceeds unity), which enables the entire discipline of fiber optics.

DIFFRACTION PRINCIPLES DERIVED FROM THE UNCERTAINTY PRINCIPLE

Subject: Optics, quantum mechanics

The Rule

Diffraction is related to the momentum uncertainty of a photon, and the basic relationship for diffraction can be easily calculated from the uncertainty principle applied to a photon.

Recall from freshman physics that the De Broglie wavelength of a particle is

$$\lambda = \frac{h}{p}$$

where λ = quantum wavelength associated with the particle
h = Planck's constant
p = momentum

and that Heisenberg's uncertainty principle states that

$$\Delta p \times \Delta d \approx h$$

where Δp = change in momentum
Δd = change or uncertainty of the position

On passing through the slit, the photon location is known to within d, so

$$\Delta p \cdot d \approx h = \lambda p \quad \text{and} \quad \frac{\Delta p}{p} \approx \frac{\lambda}{d}$$

$\Delta p / p$ defines the change in direction of the photon as a result of the encounter and is equal to the angle θ. Thus, $\theta \approx \lambda / d$.

Basis for the Rule

This is based on quantum mechanics as derived above.

Cautions and Useful Range of the Rule

Unfortunately, this does not provide the numerical constant (e.g., 2.44 for the diameter of the first Airy disk for a circular aperture). However, that constant depends on the two-dimensional shape of the aperture, and such considerations are not addressed by the above.

Usefulness of the Rule

This rule ties classic optics to quantum mechanics and underscores the importance of diffraction (and the foolishness of anyone who tries to beat it).

Also, a great way to terrify graduate students and job interviewees is to ask them to derive the basic diffraction law from the uncertainty principle. If they can, they either have great insight into the relationship of various forms or nature, were accosted by this problem before, or have a copy of this book. In any case, such a person would be worthy of a doctorate or the job.

Notes and Explanation

Fundamentally, the true nature of diffraction is the uncertainty in position of a photon rather than the messy scalar approximation derivation from Maxwell's equations. Conversely, quantum mechanics is fundamentally derived from particles being treated like waves. Remember, the Schroedinger's Equation is derived by assuming a wavelike response to a particle.

In addition, for light, diffraction is a property of the width of the light beam. It is only a coincidence that the width of the beam is sometimes determined by the width of an optic (aperture stop).

Source

Rule provided by Dr. J. Richard Kerr, 1995.

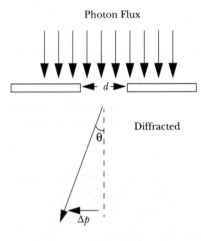

Photon Flux

d

Diffracted

θ

Δp

Figure 13.3 Photon Angle of Diffraction

GAUSSIAN APPROXIMATION TO DIFFRACTION

Subject: Optics, lasers, systems

The Rule

A Gaussian approximation to a diffraction light distribution can be found by matching the two curves at the $1/e$ point of the Gaussian. In that case, the appropriate value of sigma of the Gaussian is

$$0.431 \, \lambda \, \text{f\#} \text{ for a circular aperture}$$

$$0.358 \, \lambda \, \text{f\#} \text{ for a square aperture}$$

where f# = ratio of focal length to aperture
e = the familiar constant = 2.718

Basis for the Rule

The comparison of the Airy distribution of a circular aperture with a Gaussian provides this approximation.

Cautions and Useful Range of the Rule

The computation of encircled energy or other characteristic of the diffracted field is quite complex. This approximation allows results to be obtained that are correct to within about 10 percent.

Usefulness of the Rule

This rule is quite useful in computations where integration of the blur spot of a diffraction-limited optic is required. For example, in quadrant cell star tracers, the blur spot falls on four detectors. To compute the photon flux on each detector as a function of the position of the center of the blur, a series of integrals over the spatial extent of the detectors must be performed. This is not convenient using the exact distribution of the radiation, which includes Bessel functions. This rule provides results that are good enough in most situations.

Notes and Explanation

When the value of sigma suggested above is used to approximate a diffraction spot, one gets

$$P(r) \; = \; \frac{1}{2\pi\sigma^2} \exp\left(-\frac{r^2}{2\sigma^2} \right)$$

With this formulation, some calculations are simplified considerably as compared with trying to do the exact form derived from diffraction theory. For example, the energy distribution from a circular aperture is found to be a function made up of Bessel functions.

Source

G. Cao and X. Yu. 1994. Accuracy analysis of a Hartmenn-Shack wavefront sensor operated with a faint object. *Optical Engineering* 33(7), 2331.

QUICK ESTIMATE OF DIFFRACTION

Subject: Optics, optical performance, diffraction limit

The Rule

The angular diameter of the Airy disk is commonly used to estimate the diffraction limited spot size and defined as

$$2.44 \frac{\lambda}{D}$$

where D = aperture diameter in any linear dimensional units
  λ = wavelength in the same units as D

Or, for convenience in the infrared, this may be adjusted to:

$$D_b = 244 \frac{\lambda}{D}$$

where D_b = diffraction blur in microradians
  D = aperture diameter in centimeters
  λ = wavelength in microns

Finally, the linear diameter of the airy disk on the FPA may be estimated by:

$$2.44 \, \lambda \, (f\#)$$

Basis for the Rule

This rule results from diffraction theory.

Cautions and Useful Range of the Rule

It is important to note that this accounts for diffraction only and does not include aberrations, misalignments, scatter, and so forth.

  Both of the above relationships assume a clear circular aperture; it is different for noncircular apertures. The rule also assumes a narrow spectral band so that λ is well defined. If using a broad band of several microns (e.g., 8 to 12 μm), use the longest wavelength, because this will limit your resolution. If there is little or no energy at the longest wavelength, then you should shorten your cutoff.

  The Rayleigh criteria of diffraction limit on resolution is typically assumed to be half of this.

Usefulness of the Rule

This rule provides quick estimate of the diffraction effect and estimates on the best resolution achievable.

Notes and Explanation

It should be noted that this quick equation provides a fundamental limit. For unobstructed circular apertures, the above Airy disk will not be violated; anyone claiming better resolution must be using super-resolution and *a priori* target knowledge or synthetic apertures.

Another shortcut that is easier to remember is for visible wavelengths: $2.44 \lambda \approx$ 1000 nm or 1×10^{-6} m (which states that it takes a 1 m mirror to get a diffraction diameter of 1 μrad).

The central Airy disk is smaller than that predicted by the above rule if concentric obscurations are present, as in many reflecting telescopes. In fact, there have been telescopes designed with large central obscurations providing an aperture in the shape of an annulus. This yields a very small Airy disk. However, the fraction of energy is also less in the Airy disk; more is dumped to the outer rings, which tends to greatly reduce modulation transfer function (MTF).

For a perfect system, the bright disk (Airy disk) internal to the first dark ring contains about 84 percent of the energy; the remaining is found in low-energy rings around the Airy disk (e.g., the first bright ring surrounding this disk contains an additional 7 percent, the second has 2.8 percent and the third has 1.5 percent).

AIRY DISK DIAMETER APPROXIMATES f#

Subject: Optics, diffraction, systems

The Rule

The diameter of the Airy disk (in microns) in the visible wavelengths is approximately equal to the f# of the lens.

Basis for the Rule

This is based on diffraction theory and numerology. (Just substitute the appropriate units and you get this result.)

Cautions and Useful Range of the Rule

As stated above, the rule is valid only near wavelengths of 0.5 microns.

Usefulness of the Rule

It is useful for rapid estimates of the Airy disk diameter.

Notes and Explanation

The linear size of the first Airy dark ring is, in radius, $R = 1.22(f\lambda/D)$, where f is the focal length of the optical system. For the visible spectrum, λ is about 0.4 to 0.7 microns and f/D is a small angle approximation for the f# of the telescope. Thus, the diameter of the first Airy ring is $(2 \times 1.22 \times 0.5 \times f\#)$ microns, which is nearly equal to the numerical value of the f#.

Maximum Useful Pupil Diameter

Subject: Optics, radiometry

The Rule

The maximum effective pupil diameter (assuming f/1) is approximately limited to:

$$D \leq \frac{Ds}{\theta}$$

where D = maximum entrance pupil diameter
Ds = linear size of the detector
θ = angular instantaneous field of view in one direction from the detector pixel (sometimes called the *detector angular subtense,* or DAS)

Basis for the Rule

This rule is based on an optical system that has an f/1 speed. This is also based on geometrical optics and an approximation to the Abbe sine condition.

Cautions and Useful Range of the Rule

The rule assumes that the numerical aperture cannot exceed 1 and an f/1 cone, and that the maximum pupil decreases linearly with increasing f#. Obviously, it is possible (in some cases) to produce a system with an effective f# less than 1.

This may not strictly apply to complicated optical systems with afocals, lenslets, binary optics, and condensers.

Usefulness of the Rule

This rule is useful for estimating the effective aperture, IFOV, or detector size of a system that you do not know much about (e.g., a competitor's). This rule is useful for quickly estimating the exit pupil (and radiometric effective aperture).

Notes and Explanation

Very large FOV systems have difficulty in effectively filling their aperture with radiation that actually falls upon a FPA detector (for a given field angle). For a field of view larger than about 45°, rarely is the useful radiometric aperture the same as the physical size of the aperture.

If one assumes that the NA of a system cannot exceed 1 in air or vacuum, then by employing the Abbe sine condition and using the optical invariant, the above equation can be derived.

Example

Let's assume that you have a 256-pixel array with a 40μm detector pitch. It must subtend 45° in each dimension (that is, 45° in x and 45° in y). Your optics must therefore support a resolution of 45° (in each axis) divided by 256 elements, or 3068 μrad per pixel. If this is the case, then you can assume that your maximum useful aperture is:

$$\frac{5 \times 10^{-3}}{3.07 \times 10^{-3}} = 1.6 \text{ cm dia.}$$

The actual dimension of the first optical surface (aperture) is likely to be larger for such a wide-field system, but the entire aperture does not contribute to the energy collection for a given pixel. Again, the example assumes an f# 1 system.

MINIMUM f#

Subject: Optics, system performance, radiometry

The Rule

The theoretical minimum f# for an optical element or telescope is 0.5, the practical limit for an imaging system is about 0.7.

Basis for the Rule

Consider Fig. 13.4. When an optic is made faster than f# 0.5, a flat detector cannot respond to the rays, as they do not strike its active flat front facing surface.

The f# is defined many ways, with one of the most exact being

$$f\# = \frac{1}{2\sin\left[\tan^{-1}\left(\dfrac{D}{2FL}\right)\right]}$$

where f# = the f number

 D = effective aperture diameter (careful—this is not always the total aperture)

 FL = effective focal length from the principal surface

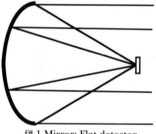

f# 1 Mirror: Flat detector can respond to all rays.

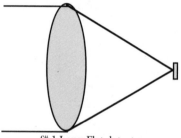

f# 1 Lens: Flat detector can respond to all rays.

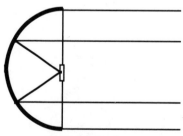

f# 0.5 Mirror: Flat detector cannot respond to rays from the edges because they do not strike an active part.

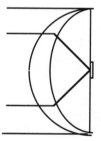

f# 0.5 Lens: Flat detector cannot respond to rays from the edges because they do not strike an active part.

Figure 13.4 f# 1 and f# 0.5 Mirrors and Lenses

Often,

$$\tan^{-1}\left(\frac{D}{2FL}\right)$$

is expressed as the angle of the rays, α.

From the above calculation, one can substitute some numbers, do the arithmetic, and see that, by definition, the minimum f# is 0.5 (the sine in never larger than 1, so f# is always larger than $1/2$).

Employing thermodynamics, one can assume that a detector radiates into its environment so as to calculate what radiates onto it. Therefore, a flat detector effectively radiates into π steradians and can only receive radiation from π steradians, indicating a minimum f# of 0.5. Using some small angle approximations, we can write the well used radiometric equation

$$A_o\Omega_o = \frac{A_d}{4\,(f\#)^2}$$

where A_0 = effective (useful) aperture
Ω_0 = solid angle (sr) subtended by the detector
A_d = area of the detector

Therefore, for the energy to be conserved, the right-hand side of the denominator must be equal to 1 or less; hence the f# must be equal to 0.5 or less. Otherwise, you could heat an object at the focal point from a colder background and devise a perpetual motion machine. Thus, the second law of thermodynamics also supports this rule as applied to optics (with some approximations).

Smith[1] points out, "The limit on the relative aperture of a well corrected optical system is that it cannot exceed twice the focal length; that is, f# 0.5 is the smallest f# attainable." In a well corrected system, the Abbe sine condition must hold. The sine condition can be expressed as

$$Y = f \sin u'$$

The limiting aperture is given by

$$f\# = \frac{f}{2Y} = \frac{f}{A} = 0.5$$

If the f# is substituted, the following theoretical limitation on the relationship of A, D, and a are obtained:

$$d_{min} = \alpha\, A \,(\text{theoretical limit})$$

where d = detector size
Y = aperture radius
f = focal length
u' = f-cone half angle

Cautions and Useful Range of the Rule

It is very difficult to make an well corrected optical system with an f# of less than 1. However, some systems where light-gathering power is important are produced with f#s in the range of 0.7, but 0.5 is the minimum (depending upon f# definition).

This assumes rectilinear projection. Oddball focal planes that are highly curved or sensitive on their sides or back can work with optics of less than a 0.5 f#. Conversely, they also effectively radiate into more than π steradians. However, most detectors (solid state or film) are flat, so this rule generally applies.

This assumes a spherical principal plane. Most are, but there are exceptions (e.g., an IR cold shield).

Usefulness of the Rule

The rule is useful for understanding the limits of optics and the limits they impose on radiometric considerations.

Notes and Explanation

This represents a practical limit on how fast you can make an optic. In general, you will always have individual optical elements and telescopes with f#s greater than 0.5. However, individual elements can be made to function with f#s of less than 0.5 (e.g., a paraboloid mirror focusing to a point, grazing incidence telescopes).

The common definition of focal length (from principal surface) divided by effective aperture is a small angle approximation.

This rule relates directly to other rules that provide support for this assertion and can be found in this chapter and Chapter 14.

Reference

1. W. Smith. 1965. Optical systems. In *Handbook of Military Infrared Technology*, ed. W. Wolfe. Washington D.C.: Office of Naval Research, 427–429.

Additional Sources

Rule provided by Dr. J. Richard Kerr, 1995.
Other information supplied by Dr. George Spencer and Max Amom, 1995.
G. Holst. 1995. *Electro-Optical Imaging System Performance*. Winter Park, FL: JCD Publishing, 459-60.
Taubkin et al. 1994. Minimum temperature difference detected by the thermal radiation of objects. *Infrared Physics and Technology* 35(5), 718.
R. Hudson. 1969. *Infrared Systems Engineering*. New York: John Wiley & Sons, 180.

f# FOR CIRCULAR OBSCURED APERTURES

Subject: Optics, systems, radiometry

The Rule

The standard definition of f# as generally applied is inappropriate when an obscuration is present. In such cases, use the effective f#:

$$f\#_{effective} = \frac{\text{effective focal length of overall system}}{\text{diameter of primary mirror}} \sqrt{\frac{1}{\left[1 - \left(\dfrac{D_o}{D_p}\right)^2\right]}}$$

where D_o = effective diameter of the obscuration
D_p = diameter of the primary mirror or other defining entrance aperture

Basis for the Rule

This is derived from basic optical theory applied to centrally (and circular) obstructed apertures.

Cautions and Useful Range of the Rule

This rule assumes that the primary mirror is the defining aperture for diffraction and assumes the object of interest (e.g., target) is at infinity.

It is valid for centered circular apertures with concentric obscurations.

Usefulness of the Rule

This relationship is wonderful for determining the f# or effective focal length that should be used and/or estimating the impact of a central obscuration.

Notes and Explanation

If there is no central obscuration, this reduces to the classic f/D_p. The effect of a central obscuration is to reduce the energy of the Airy disk although the angular extent of the bright central peak is made smaller. A central obscuration transfers energy from the Airy disk to the rings, making them more powerful. Central obscurations always tend to degrade image quality and reduce the effective f#. Even though in some point-source cases (for instance, bright binary stars) increased angular resolution may result from a central obscuration. An obscuration makes images of extended objects less acute.

There is a special-case simplification of the above rule that applies to many telescopes (especially for visible and IR astronomy). If the diameter of the central obscuration is small compared to the diameter of the aperture, then

$$\frac{1}{\sqrt{1-\varepsilon^2}} \approx 1 + \frac{1}{2}\varepsilon^2$$

where ε = obscuration diameter divided by aperture diameter = $\dfrac{D_o}{D_p}$

DEFOCUS FOR A TELESCOPE FOCUSED AT INFINITY

Subject: Optics, system design

The Rule

An optical system focused at infinity will experience a defocus when the object that it is attempting to image is not at infinity. If the object is at a finite conjugate, the angular blur of a telescope focused for infinity is:

$$\delta = D/R$$

where δ = resulting angular blur caused by misfocus from the object being closer than infinity
   D = clear aperture (aperture diameter)
   R = distance of object from the sensor's aperture

Basis For Rule

This is based on Newton's equation $(x' = -f^2/x)$, geometrical optics, and the assertion that a small a small amount of defocusing is allowable because it will not adversely affect system performance.

   Smith[1] points out that the depth of field for a system with a clear circular aperture can be expressed as

$$\frac{r}{\delta(R \pm r)} = \frac{R}{D}$$

where r = the distance of the object from the point of focus in object space (in other words, how far it is from the place where the system is focused)

Solving for r,

$$r = \frac{R^2 \delta}{(D \pm R\delta)}$$

For the image side, the above relationship can be reduced to

$$r = \frac{R^2 \delta}{D} = \frac{F^2 \delta}{D} = F\delta \, (\mathrm{f}/\#)$$

where F = focal length
   $(\mathrm{f}/\#)$ = effective f# of the system

For a hyperfocal distance, $(R + r)$ is infinity, and d is equal to D/R.

Cautions and Useful Range of the Rule

This rule is valid for systems focused at infinity in air or a vacuum but does not necessarily apply to systems in water or other media. Additionally, a fixed focus and clear circular aperture are assumed.

Usefulness of the Rule

This is useful to obtain an estimate of the minimum useful close-in range and an idea of the close-blind range of a fixed-focus seeker.

Notes and Explanation

Usually, this "defocus" is very irritating to the human observer and difficult for an image processor to handle if it exceeds about two times a pixel's angular field of view. The above relationship allows one to estimate the amount of defocus from a system, focused at infinity, that is attempting to image an object that is not at infinity. Note that the depth of field toward the optical system is smaller than that away from the system.

The above also defines the "hyperfocal" distance of a system; that is, the distance at which the system must be focused so that it remains in focus from that distance to infinity.

Reference

1. W. Smith. 1966. *Modern Optical Engineering.* New York: McGraw Hill, 133–135.

BLUR VS. FIELD-DEPENDENT ABERRATIONS

Subject: Optics, optical performance, systems

The Rules

The following expressions approximate the expected contributions to the blur diameter associated with field-dependent aberrations for various two-mirror telescope designs. The blur contribution from aberrations in radians and the off-axis angle θ are in degrees.

- Dall-Kirkham at plane of paraxial focus has a blur contribution of ≈0.001θ.
- Cassegrain at plane of paraxial focus has a blur contribution of ≈0.00062θ.
- Cassegrain with curved focal surface at best focus has a blur contribution of ≈0.00035θ.
- Ritchey-Chretien at plane of paraxial focus has a blur contribution of ≈0.0002θ.

Basis for the Rules

These relationships are derived from geometrical optics and supported by empirical experience with telescopes and ray tracing.

Cautions and Useful Range of the Rules

The analysis above assumes an f/3 optic with a primary-to-secondary axial spacing of 33 percent of the primary radius and a back focal distance 1.05 times the primary-to-secondary spacing. The real performance will depend on the optical design and some nonlinear features that are not represented in the above expressions.

The wavelength range for the above is visible only.

The above contribution to blur diameter represents the extent of a point object due to the effects of off-axis optical aberration only, not diffraction effects. Diffraction effects may dominate and should be treated as explained in other rules.

Astigmatism and field curvature often vary as the square of the field, which will cast doubt on the above approximations.

Usefulness of the Rules

For quick estimates of the blur size for various optical designs used in off-axis applications.

Notes and Explanation

There are several on-axis telescope designs named after their inventors. The difference is the aspheric curvature of the surface of the mirrors. A Dall-Kirkham has an aspheric primary and a spherical secondary; a Cassegrain has a parabolic primary and a hyperboloid secondary; a Ritchey-Chretien has a hyperboloid primary and a hyperboloid secondary.

Source

NASA Goddard Space Flight Center. 1973. *Advanced Scanners and Imaging Systems for Earth Observations.* Washington D.C.: Government Printing Office, 102, 150.

ABERRATION SCALING

Subject: Optics, aberrations, system performance

The Rule

The effect of aberrations can be scaled by field of view (FOV) and f# raised to the power of 2 or 3.

Basis for the Rule

This rule is based on geometrical optic theory. Aberrations depend on field angle and aperture size, so they can be roughly scaled by FOV and f#.

Cautions and Useful Range of the Rule

The f# and FOV should be raised to a power of 2 or 3, depending on the obscuration.

Usefulness of the Rule

This rule can be used for scaling aberrations or estimating the impact from aberrations when changing the field of view or f#.

Notes and Explanation

When someone states that the change in a field of view or f# can easily be accommodated, beware. The aberrations will change accordingly and may result in serious changes in performance. This rule is related to others in this chapter.

This rule can be stated exactly as

spherical:	$(1/f\#)^3$
coma:	$(1/f\#)^2(FOV)$
astigmatism and field curvature:	$(1/f\#)(FOV)^2$
distortion:	$(FOV)^3$

The above applies only to third-order aberrations. For residuals and higher-order aberrations, the exponents are larger.

Source

Rule partially developed by Tom Roberts, 1995.

ABERRATION DEGRADING THE BLUR SPOT

Subject: Optics, diffraction

The Rule

The angular diameter of the blur spot as a function of a given aberration can be estimated by

1. spherical aberration: $D_b = (\text{longitudinal spherical aberration})\left(\dfrac{U}{2F}\right)$

2. coma: $D_b = \dfrac{\text{coma}}{F}$

3. astigmatism: $D_b = \dfrac{1}{2}(\text{astigmatic focus difference})\dfrac{2U}{F}$

4. field curvature: $D_b = (\text{defocus})\dfrac{2U}{F}$

where D_b = diameter of the blur circle
$\quad\quad U$ = slope angle between the marginal ray and the axis at the image
$\quad\quad F$ = effective system focal length

Basis for the Rule

The above rule is based on simplifications of aberration and diffraction theory.

Cautions and Useful Range of the Rule

Although there is usually good agreement, these are estimates only. Therefore, they must be viewed with caution for large apertures or fields. This is based on third-order aberration theory.

Diffraction effects are not included and must be separately verified.

Usefulness of the Rule

The rule may be quite useful for determining if an aberration is the dominant resolution limiter and for determining aberration specifications or estimating the effect of a given aberration.

Notes and Explanation

The angular diameter (in radians) of the best focused spot may be estimated by the above equations. The angular blur, D_b, is the angle subtended from the second nodal point of the system.

When several aberrations seem to cause nearly the same (within 50 percent) blur spot diameter, then a conservative estimate of the entire blur spot can be made by summing the blurs from individual aberrations, or they may be root-sum-squared.

Source

W. Smith. 1978. Optical elements, lenses and mirrors. In *The Infrared Handbook*, ed. W. Wolfe and G. Zissis. Ann Arbor: ERIM, 9-3.

FIGURE CHANGE OF METAL MIRRORS

Subject: Optics, mirror stability

The Rule

Metal mirrors change size and figure. The figure can be expected to change about one wave per year. The higher the melting temperature, the better the stability. Additionally, *Andrade's beta law* states that the change in size of a metal is proportional to the time raised to a power, or

$$\varepsilon(t) = \beta t^m$$

where $\varepsilon(t)$ = creep strain
β = a constant dependent upon the material, stress and temperature
t = time
m = another constant (usually between 0.25 and 0.4) with a usual value of 0.33

Basis for the Rule

This is founded in metallurgical theory and empirical observations.

Cautions and Useful Range of the Rule

One wave is assumed to be at the HeNe wavelength of 632.8 nm. This is a gross approximation, so use it only as a guide.

Usefulness of the Rule

It can be used to estimate the amount the mirror is likely to change over time in both figure and size (remember, these are different parameters).

Notes and Explanation

For IR and laser applications, mirrors frequently are made from aluminum, beryllium, molybdenum alloys, or copper. Metal mirrors have many advantages and some disadvantages as compared with ceramic and glass mirrors. Among the disadvantages are cold flow, the potential for corrosion, difficulty in achieving a low scatter surface, bimetallic considerations, and long-term dimensional instability. Metal mirrors frequently change figure, especially when cycled in temperature. In general, metal mirrors can be assumed to change figure at a rate of about 1 wave per year. Pursuant to the above rule, one should allow for a change in figure of 0.6 µm per year to be conservative. Nevertheless, several fielded systems have noted a much smaller change after a year or so. Additionally, there seems to be an unproven correlation with the melting (or transition) temperature of a metal and its stability.

Sources

D. Vukobratovich. 1993. Optomechanical system design. In *Active Electro-Optical Systems*, vol. 4, ed. M. Dudzik, of *The Infrared and Electro-Optical Systems Handbook*, executive ed. J. Accetta and D. Shumaker, Ann Arbor, MI: ERIM, and Bellingham, WA: SPIE, 165–166.

E. Benn and W. Walker. 1973. Effect of microstructure on the dimensional stability of cast aluminum substrates. *Applied Optics*, vol. 12, 976–978.

L. Noethe et al. 1984. Optical wavefront analysis of thermally cycled 500 nm metallic mirrors. *Proceedings of the IAU Colloquium No. 79: Very Large Telescopes.*

F. Holden. 1964. *A Review of Dimensional Instability in Metals.* NTIS AD602379. Springfield, VA: NTIS.

C. Marshall, R. Maringer, and F. Cepollina. 1972. Dimensional stability and micromechanical properties of materials for use in an orbiting astronomical observatory. AIAA paper # 72-325. Washington, D.C.: AIAA.

RSS BLUR

Subject: Optics, systems, estimating performance

The Rule

The point spread function blur from an optical train can be estimated as:

$$R^2_{sys} = R^2_{abr} + R^2_{dif} + R^2_{def} + R^2_{jit} + R^2_{rgh} = R^2_{pix} + R^2_{ks}$$

where R_{sys} = radius of the system point spread function

   R_{abr} = radius of the blur due to geometric aberrations excluding defocus

   R_{dif} = radius of the blur due to diffraction

   R_{def} = radius of the blur due to defocus alone (when aberrations are small—
      if aberrations are large, then the defocus effects should be another
      part of R_{abr})

   R_{jit} = radius of the blur due to jitter or the integrated image motion during
      the effective exposure (integration) time

   R_{rgh} = radius of the blur due to surface defects (or roughness) that cause
      near forward scattering

   R_{pix} = "radius" or equivalent measure of the detector pixel width

   R_{ks} = radius of any other blur contributor (usually, you should throw into
      this term everything but the kitchen sink)

Basis for the Rule

All the blurs have at least a vaguely Gaussian profiles of the form $\exp[-(r/R)^2]$. Therefore, the total system blur profile is approximately the convolution of the contributing blur profiles.

Cautions and Useful Range of the Rule

The appropriate radius is partly a matter of judgment. For example, what is the best fit $1/e$ width of a Gaussian approximation to a "top-hat" function (such as is the case with the FPA pixel width)? Any reasonable fit will serve. Moreover, when there are many contributions, the system point spread function will tend to be Gaussian, even if none of the contributors are. The sum-of-the-squares process tends to emphasize the influence of one or a few large contributors; therefore, small contributors can be identified and ignored.

The electronics, processing, and display may also degrade the overall system point spread function and should also be root-sum-squared if significant.

Usefulness of the Rule

This rule

- is easy to calculate
- can be applied quickly
- gives an indication of dominating contributors
- gives an indication of insignificant contributors that can be ignored
- gives an indication of where the time and money should be spent to increase (or restore) performance
- allows a system-wide balancing of errors

Notes and Explanation

This rule is similar to the more simplistic "Optical Performance of a Telescope" rule (page 278).

Source

Rule provided by W. M. Bloomquist, 1995.

RELAYS EFFECT ON FOV

Subject: Optics, systems

The Rule

A sequence of relays will result in a small practical field of view and/or poor image quality at the edge of the field.

Basis for the Rule

This rule is based on general geometric optics. This effect occurs because relays result in poor image quality at the edge of the field and, generally, the field is reduced to eliminate the poor image regions. Relays all have field curvature. The field curvature of successive relays will add. Add many of them, and the resulting field curvature will limit the usable image to something small.

Cautions and Useful Range of the Rule

The above rule points out a truism, although sometimes a small FOV is all that is required. Also be wary of the relay's effect on beam diameters.

Usefulness of the Rule

This rule is useful as a reminder that there is a down side to a series of relays.

Notes and Explanation

Often, the detector/dewar apparatus needs to be packaged far from the aperture. This unfortunate condition usually requires the optics designer to string fiber optic bundles or relay the image through many turns and bends. When more than a couple relays are cascaded together (e.g., when directing a beam through a gimbal), the relays may limit the field of view, unless the field lenses are used to relay the pupil. This is common knowledge with periscope designers, who have battled this outcome for a century.

Source

Rule provided by W. M. Bloomquist, 1995.

THICKNESS-TO-DIAMETER RATIO FOR MIRRORS

Subject: Optics, shop optics, mirrors, optomechanics

The Rule

Hudson[1] reports that a rule of thumb is to make a glass mirror's thickness at least 1/6 of its diameter, and Roberts[2] points out that the ratio for aluminum can be 1:8 to 1:10, and 1:12 for beryllium and advanced composites.

Basis for the Rule

This rule is based on empirical observations, geometry, and material properties. The thickness of the mirror must be sufficient to support the mirror's shape. The flexure of a circular disk supported at the edge is proportional to the fourth power of the diameter and the −2 power of the thickness.

Cautions and Useful Range of the Rule

The above assertion does not account for advanced, lightweight techniques and is very material dependent.

This assumes a 1 g environment with a mirror supported on the edge only. For application of more than 1 g (or less), the ratio can be adjusted accordingly.

Small mirrors often can be much thinner—1/20th or less.

Usefulness of the Rule

This rule allows a quick estimate of the expected thickness of a mirror.

Notes and Explanation

The stiffness of a mirror is dependent on the material, mounting, and thickness. For most applications, a mirror will be able to retain its figure in a gravity environment when supported on the edges if it is of the above thickness.

References

1. Rule partially developed by Tom Roberts, 1995.
2. R. Hudson. 1969. *Infrared Systems Engineering*. New York: John Wiley & Sons, 201.

MASS IS PROPORTIONAL TO ELEMENT SIZE CUBED

Subject: Optics, telescope mass, optomechanics

The Rule

The difference in mass of two optical elements (or telescope assemblies) of different sizes is usually proportional to the ratio of their diameters raised to a power between 2 and 3. Mathematically,

$$M_{o1} = M_{o2}\left(\frac{D_{o1}^n}{D_{o2}^n}\right)$$

where M_{o1} = unknown mass of optic or telescope
   M_{o2} = known mass of similar optic or telescope
   D_{o1} = diameter of unknown-mass optic or telescope aperture
   D_{o2} = diameter of known-mass optic or telescope aperture
   n = a constant between 2 and 3—usually 2.7

Basis for the Rule

This rule is based on empirical observations of the present state of the art. An optical element of the same material and figure will generally need to be thicker as its diameter is scaled (to maintain a given surface), so the volume follows the same rule.

Cautions and Useful Ranges of the Rule

To use this rule, the telescopes' diameters should be within a factor of 3 of each other, the optics must be of the same type and material (e.g., two on -axis reflective Cassegrains made of aluminum), the optics should be of the same prescription (or close), mechanical and environmental specifications must be the similar, and off-axis stray light rejection specifications should be comparable.

This rule is commonly valid for optics from 1 cm to several meters in diameter.

Usefulness of the Rule

Knowing this rule can help with system trade-offs of comparing the mass impact of changing the optics size and estimating if a given optic requires advanced light-weighting techniques (and, hence, lots of bucks) to meet mass goals.

Notes and Explanation

The mass of a given mirror or lens depends on its material, density, volume, optical prescription, required strength, and the lightweighting techniques applied. An estimate of the mass of an unknown optical element can be made, based on the known mass of a similar element. Telescope mass usually tracks the mass of the optical elements linearly, so this rule can also hold for entire telescope assemblies.

Generally, the diameter-to-thickness ratio for normal optical elements is 6:1 to 10:1.

Vukobratovich[1] suggests that the exponent n is equal to 2.92. He states that state-of-the-art lightweight mirror mass can be estimated by $W = 53D^{2.67}$ where W is the mass in kilograms, and D is the diameter in meters.

Example

One wishes to know the mass impact of increasing the aperture diameter from 10 cm to 30 cm—let's say a 10 cm diameter silicon carbide (SiC) optic mirror of a special lightweight design that weighs 20 grams to use for scaling. Therefore, the desired 30 cm mirror would weigh approximately $(20)(30)^{2.7}$ divided by $(10)^{2.7}$, or 388 g. Thus, one can estimate the mass of the larger mirror to be about 400 g.

Reference

1. D. Vukobratovich. 1993. Optomechanical system design. In *Active Electro-Optical Systems*, vol. 4, ed. M. Dudzik, of *The Infrared and Electro-Optical Systems Handbook*, executive ed. J. Accetta and D. Shumaker, Ann Arbor, MI: ERIM, and Bellingham, WA: SPIE, 159–160.

Additional Source

J. Miller. 1994. *Principles of Infrared Technology.* New York: Van Nostrand Reinhold, 88–91.

LARGEST OPTICAL ELEMENT DRIVES THE MASS OF THE TELESCOPE

Subject: Optics, shop optics, telescope assembly mass

The Rule

The mass of a telescope assembly is directly proportional to the mass of the largest optical element.

Basis for the Rule

This is based on empirical observations and the state of the art.

Cautions and Useful Ranges of the Rule

If used as a comparison, optical element masses should be within a factor of 3 of each other. Telescopes must be of the same type and material (e.g., two on-axis, reflective Cassegrains made of aluminum). Usually this is valid for telescopes ranging from a few centimeters to a meter or two in aperture. Telescopes should have the same number of elements and have similar environmental stability, stiffness, and slewing specifications to apply this rule. Finally, the off-axis stray-light rejection specifications should be comparable.

Usefulness of the Rule

The rule is useful (with the other rules on optics and telescope mass) for system-trade-offs when comparing the mass impact of changing the optics size and estimating if a given telescope requires advanced technologies to meet the mass goals.

Notes and Explanation

A given telescope's mass depends on its design, composition materials, size, required strength, and the lightweighting techniques applied. An estimation of the mass of an unknown telescope assembly can be obtained by scaling based on the known mass of a similar element. Telescope assembly masses usually track the mass of the heaviest optical elements linearly.

One should be sure to use the heaviest optic element. Usually, this is the largest, but it may not be in so in refractive designs with thick lenses. The largest and heaviest element is usually the first objective or the primary mirror. However, in off-axis reflective telescopes, and some wide field of view designs, the largest element is not the primary or objective. If the telescope is a Schmidt, the primary is larger than the clear aperture or the correcting plate.

OPTICAL ELEMENT COST

Subject: Optics, cost, systems

The Rule

An optical element's cost (or the cost of a telescope) is proportional to its diameter raised to a power, divided by the wavelength of operation of the mirror raised to another power and the f# raised to yet another power.

$$\text{Cost} \propto \frac{D^n}{\lambda^m (\text{f\#})^x}$$

where D = diameter of the largest optical element
n = an adjustment constant of ≈ 2.7 (usually $2 < n < 3$)
λ = bandpass cut-on (since this is relative, it can be in any desired units)
m = another adjustment constant of ≈ 2 (usually $0.8 < m < 4$)
f# = f number
x = yet another adjustment constant ($x \approx 0.5$ ($0.1 < x < 1.5$) for f# $>\approx 2.5$, $x \approx 2$ ($1 < x < 10$) for f# $<\approx 2.5$)

Basis for the Rule

This rule is based on empirical curve fitting of current state-of-the-art custom optics. Generally, the larger the diameter, the more an optic will cost, and this is a very dramatic increase. Also, the lower the bandpass cut-on, the more accurate the figure needs to be ground, and the less surface roughness so, again, the optic will cost more. Lastly, faster f numbers require higher curvatures, so they are more difficult to produce and have lower yields. Again, this means higher cost.

Cautions and Useful Range of the Rule

This rule provides crude approximations for scaling similar optics of the same material (cost is a strong function of the difficulty in working with certain materials). The rule does not account for costs associated with testing, tolerances, and fields of view.

This rule should not be used to compare systems unless all have D, λ, and f# in similar ranges. The rule does not account for nontechnical cost drivers such as schedule, how hungry the vendor is, the vendor's past experience with similar optics, and so on.

Usefulness of the Rule

The above is useful for a first-cut quick estimate or a scaled comparison of the cost of optics and telescopes. It also serves to illustrate the typical cost drivers.

Notes and Explanation

Folklore and "wive's tales" indicate that a given optical element's cost is somewhat proportional to the diameter raised to a power divided by the surface quality (figure and roughness). This also applies to radio telescopes, antennas, and lots of high-tech things. The folklore can be refined to include the effects of the speed. The larger the f#, the easier it is to make optics of a given quality. Additionally,

Hudson[1] reports that, for large f numbers, parabolic mirrors are not significantly more expensive than spheres, but for optics with an f number less than 3, parabolas are more expensive. For the same f number, mirror costs increase somewhat more quickly than the square of the increase in diameter. The cost of an optic is somewhat linear (for small differences) to the final roughness and figure. The scatter from an optical surface varies by the roughness squared and $(1/\lambda)^2$ (see the associated scatter rule). Therefore, the wavelength, which is usually more readily known, can be substituted.

Reference

1. R. Hudson. 1969. *Infrared Systems Engineering*. New York: John Wiley & Sons, 201.

Other Source

J. Miller. 1994. *Principles of Infrared Technology*. New York: Van Nostrand Reinhold, 93.

MIRROR SUPPORT CRITERIA

Subject: Optics, optomechanics, shop optics, systems

The Rule

The minimum number of mirror support points needed to control self-weight deflections of an optical element as a function of the permissible peak-to-valley deformation is:

$$N = \left(\frac{1.5\,r^2}{t} \right) \left(\frac{\rho}{E\delta} \right)^{1/2}$$

where N = minimum number of support points
  r = mirror radius (mm)
  E = mirror modulus of elasticity (e.g., kg/mm^2)
  δ = allowable peak-to-valley distortion in linear measurements (mm)
  ρ = mirror material density (kg/mm^3)
  t = mirror thickness (mm)

Basis for the Rule

This rule results from empirical observations and logic.

Cautions and Useful Range of the Rule

Yoder[1] reports that this has proved satisfactory for mirrors ranging from 1 m in diameter and 10 cm thick to 2.6 m in diameter and 30 cm thick.

Usefulness of the Rule

The rule can be very useful in determining the size and design of a test fixture for mirror support, the size and design of a mirror mount, and the number of supports for a polishing fixture (must include the extra weight of the polishing tools and auxiliary weights).

Notes and Explanation

It is always best to support a mirror at as many points as possible. However, design constraints frequently lead to simple and few supports. Hall[2] derived the above relationship to estimate the number of support points needed to prevent self-weight deflections larger than a specific peak-to-valley allowable deflection. Yoder[1] gives the origin of this as Hall[2].

Example

Consider a stiff, thick, LWIR mirror of 1 m diameter: r = mirror radius of 500 mm, E = modulus of elasticity 7.6×10^5 kgf/mm^2, δ = allowable peak-to-valley distortion of 1 micron or 10^{-3} mm, ρ = mirror material density of 2.7×10^{-3} kg/mm^3, t = mirror thickness of 100 mm.

$$N = \left(\frac{1.5\,r^2}{t} \right) \left(\frac{\rho}{E\delta} \right)^{1/2} = \frac{(1.5)\,(500)^2}{100} \sqrt{\frac{2.7\times10^{-3}}{(7.6\times10^5)\,(10^{-3})}} = 2.2$$

Therefore, three supports should provide proper mounting for this mirror.

References

1. P. Yoder. 1986. *Opto-Mechanical Systems Design.* New York: Marcel Dekker, 310.
2. H. Hall. 1970. Problems in Adapting small mirror fabrication techniques to large mirrors. In *Proceedings of the Optical Telescope Technology Workshop.* NASA report Sp-233, 149.

LIMIT ON FOV FOR REFLECTIVE TELESCOPES

Subject: Optics, telescope design

The Rule

An all-reflective (mirror) telescope can have a field of view of no more than about 10 (or maybe 20) square degrees (°°) before heroic measures are required to try to make it work, such as curved focal planes, extra mirrors, and super-expensive compound curvature mirrors.

Basis for the Rule

This is based on the limitations imposed by geometrical optics on fields of view coupled with current optical manufacturing technology.

Cautions and Useful Range of the Rule

Because this is a generalized rant by the authors, cautions must be applied.

Some designs utilizing cylindrical optics may exceed this limit and remain producible.

Schmidt telescopes frequently have fields of view approaching or exceeding $20°°$. However, they do this with a refractive *corrector plate* as the first optical element and employ curved focal planes.

Usefulness of the Rule

This is valuable for underscoring the fact that reflective telescopes are narrow field of view devices, and also as a sanity check on a wide-field-of-view (WFOV) reflective design.

Notes and Explanation

The *square degrees* unit used herein refers to the product of the two angular dimensions of the field of view. Often, you must use all-reflecting optics and tolerate a small field of view. Do not whine about this. It does not occur just to make your job harder; it is dictated by geometry and physics.

ALIGNMENT PROBLEMS

Subject: Optics, systems, shop optics

The Rule

Alignment is always better in the lab than in the field. The one exception seems to be when it is worse in the field than it is in the lab.

Basis for the Rule

This is based on empirical observations.

Cautions and Useful Range of the Rule

A properly designed and toleranced optical system can have slight alignment changes over the life of a system, which may not noticeably affect performance.

Usefulness of the Rule

This reminds us to leave some margin for alignment in optics tolerancing and radiometric performance calculations. An optical design that only meets its performance requirements when precisely aligned will simply never meet its requirements in the real world.

Notes and Explanation

Aligning a large train of optical elements is a black art. If the gods smile on you, you may achieve an alignment in your lab suitable for optimum performance. The performance is always made worse by misalignment. However, it is wise to allow for some degradation when the system is really put to use. Generally, alignment is less perfect in orbit than it was on the ground. Temperature gradients cause misalignment. Zero gravity allows misalignment. Launch vibration and shock cause misalignment. Unstable optical and structural materials cause misalignment. These materials are usually the low-mass materials that are attractive for space applications. Thermal cycling causes misalignment. Fighters on afterburner tend to cause misalignment. Landings on aircraft carriers tend to cause misalignments. Dropping the optical trains on the ground tends to cause misalignment. Spacecraft, when alternatively exposed to the sun and darkness, tend to have misalignments. The list goes on and on . . . .

INDEX IMPACTS ON WFE

Subject: Optics, system performance, wavefront error

The Rule

Wavefront error can be calculated as follows:

$$WFE = \sigma|\Delta n|$$

where *WFE* = wavefront error in wavelengths of light at a given wavelength (The HeNe wavelength of 0.63 μm is typically used for optics in the visible.)

σ = surface error in wavelengths of light

Δn = magnitude of the change in the index of refraction across the interface ($n_1 - n_2$)

Basis for the Rule

This equation is a result of geometrical optics, wavefront calculations, and basic physics.

Cautions and Useful Range of the Rule

This rule is valid across the electromagnetic spectrum.

Usefulness of the Rule

This simple rule allows one to estimate total wavefront error.

It provides guidance in determining a specification for the irregularity (flatness) of an optical surface based on the material's index of refraction and the allowable wavefront error.

The rule allows cost and producibility trade-offs based on the difficulty in achieving a given surface figure and the material's index. This gives the project engineer important ammunition for convincing the optic designer to change to a lower index material (or vice versa).

Notes and Explanation

When estimating overall system performance, the quantity that the analyst really desires is the wavefront error, not the surface error. The wavefront error can be related to the surface error by the above rule. We can see that the more an optical material's index varies from that of 1 (the index of refraction for a vacuum, and approximately that of air), the more stringent the surface figure needs to be.

IR materials tend to be the worst due to their high index of refraction ($n_1 - n_2$ being on the order of 2 to 4.5). However, IR wavelengths are also long, so the absolute tolerance on *WFE* is not as stringent, and this offsets many of the manufacturability headaches. Mirrors tend to be the most critical, as their change in index is $n_1 - n_2 = 1 - (-1) \approx +2$. Fortunately for the pocketbook of anyone who wears eyeglasses, ordinary glass is less critical: $n_1 - n_2 \approx 1 - (1.5) = -0.5$. Immersed or buried surfaces hardly matter, as $n_1 - n_2 \approx 1.6 - (1.5) = -0.1$.

Source

Rule provided by W. M. Bloomquist, 1995.

ANTIREFLECTION COATING INDEX

Subject: Optics, coatings

The Rule

Antireflection coatings have minimum reflectance when their index of refraction is the square root of the index of refraction of the substrate, and they are a quarter of a wavelength thick.

Basis for the Rule

This rule is based on Fresnel reflection coefficients (which can be derived from Maxwell's equations, if you must) and wavefront, interference theory, and basic ray tracing.

Cautions and Useful Range of the Rule

Benefits are gained for antireflection coatings even if the above is only partially met.

Usefulness of the Rule

This is useful in coating design and underscores the nature of antireflection coatings.

Notes and Explanation

Antireflection coatings should be applied to the surface of optics and focal planes. For high index of refraction materials (such as germanium), such coatings can almost double the throughput. If the above rule is met, the first surface reflection is reduced to a minimum at a given wavelength.

AXIAL PLACEMENT OF THE FOCAL PLANE

Subject: Optics, system design, optomechanics

The Rule

The axial placement (focusing) accuracy along the optic (commonly referred to as the Z) axis needed for the focal plane (detector or film) is approximately $5\lambda f\#^2$.

Basis of the Rule

Geometrical optics coupled with an allowance for slight defocus yields this rule.

Cautions and Useful Range of the Rule

It is defined as a $\pm$, so use one-half in any direction and allow defocusing to degrade resolution by 1.4 times.

This rule assumes narrow field of view systems and 1/4-wave figures on the optics.

Usefulness of the Rule

This can be used effectively in determining the tolerance requirement to the location of the FPA, estimating the manufacturability and producibility of a system, and in setting the requirements for athermalization and vibration.

Notes and Explanation

There is a finite amount of error allowed for the placement of the focal plane with respect to the optics. As the placement error grows, a defocusing effect will increase. This rule of thumb is a generalization that allows the defocus to grow to about the size of the Airy disk. Usually, one can repeatedly and easily place an FPA to 0.05 mm, which is good enough if $\lambda \approx 0.5$ micron and the f# is about 4 or more.

CONVENIENT APPROXIMATIONS

Subject: Optics, radiometry, shop optics, telescopes, microscopes

The Rules

Levi[1] gives the following rules:

1. When an f/8 lens forms an image of a distant object, the image illumination is about 1/100 of the object luminance.

2. The power of a telescope should not exceed the radius (in millimeters) of the clear aperture of its objective.

3. The maximum useful magnification of a microscope is 1000 times the numerical aperture of the objective.

4. When the index of refraction is 1.5, the focal length of an equiconvex thin lens equals the radius of curvature of its faces. The focal length of a plano-convex lens is twice this value

Basis for the Rules

It appears that Levi based these rules on simplifications and approximations of lens design, diffraction theory, and empirical and practical observations.

Cautions and Useful Range of the Rules

These rules are generally applicable for normal image-forming systems and are most applicable to visible bands for use with the human eye.

Usefulness of the Rules

These rules represent simple checks to facilitate the estimation of lens performance and to conduct a sanity check on various designs. Additionally, they represent a good places from which to start scaling.

Notes and Explanation

When the magnification for a telescope or microscope greatly exceeds these levels, it exceeds the resolving power of the eye. Therefore, additional magnification only allows a more detailed view of the Airy disk (which is quite boring and not "useful").

Amateur astronomers often use the diameter of the telescope in millimeters as opposed to the radius as Levi suggests.

Reference

1. L. Levi. 1968. *Applied Optics*, vol. 1, *A Guide to Optical System Design*. New York: John Wiley & Sons, 471, 475, 487–488.

Dome Collapse Pressure

Subject: Optics, domes, windows

The Rule

An optical dome will collapse when the pressure equals

$$P_{collapse} = \frac{0.8E}{\sqrt{1-v^2}}\left(\frac{R_o - R_i}{R_o}\right)^2$$

where $P_{collapse}$ = pressure at which the dome will collapse
$\quad\quad E$ = elastic modulus of the material of the dome
$\quad\quad v$ = Poisson ratio for the material
$\quad\quad R_o$ = outer radius
$\quad\quad R_i$ = inner radius

If the pressure loading is uniform (such as with aerodynamic loading) on the projected area, this simplifies to (for thin domes):

$$\text{Stress} = \frac{PR}{2h}$$

where P = magnitude of the uniform loading
$\quad\quad R$ = radius of the dome
$\quad\quad h$ = thickness of the dome

Basis for the Rule

This rule is empirically developed from extensive model testing and is a simplification of more complex equations.

Cautions and Useful Range of the Rule

Such a collapse by elastic buckling does not account for damage due to impacts (e.g., dragonflies).
    A safety factor should also be applied.
    This is valid for domes of 180° or less.

Usefulness of the Rule

It can be used for estimating the thickness required to survive a pressure differential or estimating a pressure differential that will break a given dome.

Notes and Explanation

Frequently, sensors require a dome or window through which they can view. This is usually true for sensors placed on tactical air platforms, or underwater, and for missile seekers. Domes and windows are sometimes employed in space for protection. A dome will buckle when the pressure equals the above equation. When properly designed and seated, domes can survive extreme pressures. One seeker's dome routinely survives 11,000 g and 26 Mpa.

Source

D. Vukobratovich. 1993. Optomechanical system design. In *Active Electro-Optical Systems*, vol. 4, ed. M. Dudzik, of *The Infrared and Electro-Optical Systems Handbook*, executive ed. J. Accetta and D. Shumaker, Ann Arbor, MI: ERIM, and Bellingham, WA: SPIE, 127.

PRESSURE ON A PLANE WINDOW

Subject: Optics, windows, pressure deformation, survivability

The Rule

To survive a pressure difference, a plane window should have the following minimum thickness:

$$t = \frac{d}{2\sqrt{\left[\dfrac{8\sigma_f}{3\Delta P(x+\upsilon)\,SF}\right]}}$$

where t = necessary minimum thickness of the window
  d = diameter of the window
  σ_f = fracture stress of the window material
  ΔP = axial pressure differential on the window
   x = a constant depending upon the mounting for the window (Vukobratovich[1] suggests 3 for a simply supported window and 1 for a clamped window)
   υ = Poisson's ratio for the window material
  SF = a safety factor

Basis for the Rule

This rule is based on pressure dynamics, backed up by measurements. It is based on the mechanical engineering assertion that tensile stress should not exceed the fracture stress of the material divided by a safety factor.

Cautions and Useful Range of the Rule

This applies only to flat (unpowered) circular windows with mountings at the edge. Different shaped windows require different forms of the equation.

This rule assumes brittle materials, as most windows are.

A safety factor should be applied, and large safety factors are often used (e.g., 10 or 20).

This rule assumes a static pressure differential loading. When stressed by such a pressure differential, the window may not break, but its survival will be sensitive to a momentary increase in pressure (as a dragonfly hitting the surface) and may break.

Usefulness of the Rule

This rule is useful for determining the required thickness of the window during design or determining the maximum pressure differential that a given window can survive.

Notes and Explanation

A window's thickness is usually driven by a desire that it survive minor impacts (e.g., bugs, micrometeorites, or fish) or a pressure differential load across its surface. When a pressure is applied to a window that is supported at its edges, the center bows inward, forming a weak negative meniscus lens. If the pressure is strong enough, the bowing will be great enough to break the window.

The equation allows a good estimate for the thickness based on constant pressure loading (e.g., aerodynamic and underwater loads). However, for high optical quality instruments, the distortion from the bending of the window (into a meniscus form) may be the driving constraint. The following equation estimates the expected optical path difference for a window undergoing uniform loading from a difference in pressure. Typically, the allowable OPD depends on the wavelength and modulation transfer function (MTF) of the system. It is usually a few "waves."

$$OPD = 8.89 \times 10^{-3} \left[\frac{(n-1)\Delta P^2 d^6}{E^2 t^5} \right]$$

where OPD = maximum optical path difference
n = index of refraction of the window material
ΔP = axial pressure differential across the window
d = window diameter
E = elastic modulus of the material
t = window thickness

Reference

D. Vukobratovich. 1993. Optomechanical system design. In *Active Electro-Optical Systems*, vol. 4, ed. M. Dudzik, of *The Infrared and Electro-Optical Systems Handbook*, executive ed. J. Accetta and D. Shumaker, Ann Arbor, MI: ERIM, and Bellingham, WA: SPIE, 123–125.

SCATTER DEPENDS ON SURFACE ROUGHNESS AND WAVELENGTH

Subject: Optics, scatter, surface physics

The Rule

Scatter depends on the surface roughness and wavelength in the following form:

$$TIS \propto \left(\frac{4\pi\sigma\cos\theta}{\lambda} \right)^2$$

where TIS = total integrated scatter
π = pi (you know, 3.14159 ....)
σ = RMS surface roughness
θ = angle of the incident light to be scattered
λ = wavelength

Clearly, σ and λ must have the same units.

Basis for the Rule

This originated through studies of radar scattering off ocean waves, but it has been verified to be accurate when applied to optics.

Cautions and Useful Range of the Rule

This rule assumes a smooth, clean surface that is large compared to the wavelength (all attributes of a typical optical surface). It also assumes that the height distribution of the roughness is Gaussian and that the surface is much wider than the correlation distance of the roughness.

Usefulness of the Rule

This is useful for estimating the amount of scatter from an optical surface and scaling the scatter between different wavelengths or optical surfaces of different surface roughnesses.

It is also useful as a reminder of the properties that result from (or affect) the level of scatter.

Notes and Explanation

When light is refracted, diffracted, or reflected from an optic, some (hopefully small) portion will scatter. That is, some of the rays will sent on an unintended course. The scatter is related to the surface roughness and cleanliness. Shorter wavelengths will always tend to scatter more than longer wavelengths (which is why the sky is blue). This scatter can be a performance driver for systems that must operate near bright sources, and a cost driver for low-scatter, short-wavelength systems.

Source

J. Stover. 1990. *Optical Scattering.* New York: McGraw-Hill, 17–19.

Package vs. Focal Length

Subject: Optics, telescope design, cost, systems

The Rule

Decreasing the ratio of mechanical length to optical focal length requires shortening the focal length of the individual components and usually results in larger aberrations and more stringent manufacturing tolerances.

Basis for the Rule

This comes from empirical observations, the scaling of aberrations and geometrical optics, and the practical problems in packaging long focal lengths in small packages.

Cautions and Useful Range of the Rule

Remember that this is a gross generalization only, so caution should always be used. Clearly, this does not account for the use of binary optics or other Herculean efforts.

Usefulness of the Rule

The above rule is important to consider when changing specifications, in underscoring the resulting effects of a change in exterior package size, and as a general caution to consider during design.

Notes and Explanation

If someone flippantly states that a decrease in the optics package size can be accommodated easily, be cautious. The total dimensions and weight of a telescope usually are strong drivers on the design and cost. Usually, decreasing the telescope's physical package will reduce performance and producibility while increasing cost and development time.

PEAK TO VALLEY APPROXIMATES FOUR TIMES THE RMS

Subject: Optics, analysis, relation between figure and wavefront error

The Rule

The peak-to-valley error of the surface figure of an optical element is four times the RMS wavefront error.

Basis for the Rule

The rule is derived from an analysis of simple aberrations and various high-spatial-frequency wavefront errors. Mahajan[1] gives the Zernike polynomials in a form such that they all have unit RMS value and peak-to-valley (PV) ratios that vary. This allows the derivation of the ratios, and 4 is an approximately good median.

Cautions and Useful Range of the Rule

The choice of "4" is arbitrary, as specific aberrations have slightly different ratios for the PV error to RMS (ranging from $\sqrt{3}$ to 8).

1. Lower-order (Seidel aberrations) have ratios from $\sqrt{3}$ to $\sqrt{32}$.
2. Crinky wavefronts (more or less random) have a ratio of $2\sqrt{2}$.

This rule also seems to hold for many types of nonstandard aberrations such as

1. circumferential groove in the wavefront that varies sinusoidally with radius,
2. grooves varying cosinusoidally
3. grooves with a square wave variation with radius
4. a two-level zone plate

In each of the four aberrations, the PV to RMS ratio is either 2 or $2\sqrt{2}$ when there are many periods from center to edge. For higher-order Zernikes, the ratio seems to tend toward 8. Truly random wavefront errors are arguably statistically something like the many groove or higher-order Zernike and therefore have PV to RMS ratios somewhere between 2 and 8. Four is between 2 and 8 (as is, say, 5), so 4 was chosen as a reasonable (e.g., never quite right) approximation.

Usefulness of the Rule

This rule allows a common-ground comparison of optics specified in different ways.

Often, the PV figure error is available by casual inspection of an interferogram. Conversion to the RMS error is more tedious than using this rule, since it requires some computation of a two-dimensional integral.

RMS wavefront error is useful in computing other quality measures (e.g., Strehl ratio) as long as it is not too large.

Notes and Explanation

Generally, the RMS wavefront error is about 1/4 of the PV error.

Any wavefront that can be described from either analysis or direct measurement can be characterized by a PV or RMS deviation from its ideal figure. RMS involves integration over the whole wavefront, so it draws on more information about the wavefront (although it yields a single number). Any useful wavefront probably does not have pathologies such as discontinuities, so it is in some sense smooth. In turn, the smoothness implies that PV and RMS measures are more or less proportional.

Some specific cases are shown in the following table:

Type of aberration	PV/RMS
Defocus	1.73
Primary spherical	5.66
Astigmatism	4.90
Secondary spherical	2.65
Phonograph record	2.83
When in doubt use	4

Reference

1. V. Mahajan. 1994. Zernike polynomials and optical aberrations of systems with circular pupils. *Applied Optics,* vol. 33, 8121–8124.

Source

Rule provided by W. M. Bloomquist, 1995.

OPTICAL PERFORMANCE OF A TELESCOPE

Subject: Optics, performance

The Rule

A blur circle's angular diameter may be approximated by

$$\theta = \sqrt{\theta_D^2 + \theta_F^2 + \theta_A^2}$$

where θ_D = diffraction effect
θ_F = figure imperfection effect
θ_A = misalignment of optics effect

A typical expression for θ_D is $3.69(\lambda/D)$ for a system with a central obscuration radius of 0.4 of the aperture radius.

$$\theta_F = 16\frac{x}{D}$$

A typical value of x might be about $(\lambda/3)$ or $(\lambda/4)$. This value again assumes the obscuration mentioned above.

The effect of misalignment is approximately

$$\theta_A \approx \frac{mD\Delta}{k_1 \, EFL^2}$$

where D = diameter of the optical system
λ = wavelength of operation (or cutoff of a wide bandpass)
x = optics figure
m = magnification of the telescope
Δ = lateral misalignment
k_1 = a constant depending on the specific optical design
EFL = effective focal length of the telescope

Basis for the Rule

This rule is based on approximation of diffraction theory and general experience with telescopes and ray tracing.

Cautions and Useful Range of the Rule

The real performance will depend on the optical design and some nonlinear features that are not represented in the equations.

The expression for θ_D assumes a reflective telescope with a 40 percent obscuration and should be adjusted for other obscuration ratios.

Usefulness of the Rule

The rule provides a quick estimate of the blur size that can be achieved for a given level of optical quality and alignment effort.

The rule can be used for determining optical requirements.

Notes and Explanation

It is important to understand that these imperfections in optics sum via root-sum-squared (RSS). Having an element with exceptional figure does not provide a smaller spot if the alignment cannot be done properly.

This gives a blur circle roughly equivalent to the Airy disk diameter, which contains about 80 percent of the energy entering a telescope that has a secondary-to-primary obscuration ratio of 40 percent.

Source

NASA Goddard Space Flight Center. 1973. *Advanced Scanners and Imaging Systems for Earth Observations.* Washington DC: Government Printing Office, 99–101.

LINEAR APPROXIMATION FOR MTF

Subject: Optics, systems, MTF

The Rule

The MTF of a system can be estimated by:

$$MTF_{diff}(f) \approx 1 - 1.3\left(\frac{f}{f_c}\right)$$

where MTF_{diff} = MTF for a diffraction-limited system
  f_c = spatial frequency cutoff
  f = spatial frequency in question

Alternatively, in the visible, the MTF of a perfectly circular lens can be approximated by:

$$MTF \approx 1 - \frac{f\#\upsilon}{1500} \text{ or } 1 - \frac{\upsilon^*}{3A}$$

where MTF = modulation transfer function
  f# = f# of the lens
  υ = spatial frequency in lines per mm
  υ^* = spatial frequency in lines per micron
  A = numerical aperture

Basis for the Rule

This rule results from a linear approximation to an actual diffraction-limited MTF and approximation of diffraction theory and curve fitting MTF calculations.

Cautions and Useful Range of the Rule

The rule assumes a diffraction-limited circular aperture that limits the MTF. Sometimes, MTF is limited by components other than the optics (e.g., the detector or focal length). It also assumes a single wavelength, but the relationship tends to hold pretty well for narrow bandpasses.

The ratio of f to f_c must be less than 0.7; this linear approximation diverges substantially when the ratio is larger.

As the rule is for a perfect lens, degrade it for real-world optics.

The rule is valid for an MTF greater than 0.15 and incoherent illumination.

Usefulness of the Rule

It is useful for making rapid estimates of the upper bounds of an MTF.

Notes and Explanation

A spatial frequency as sampled on a focal plane is a function of f#. However, the maximum spatial frequency (best resolution) is determined by aperture and wavelength for diffraction-limited systems (sampling the Airy disk more than a few times is of no practical use).

When rigorously calculated, the MTF seems to vary approximately linearly for diffraction-limited systems when the spatial resolution exceeds that of the object of interest. The rule provides excellent correlation for ratios of 0.4 or less. When the object (or line pair) of interest approaches the diffraction limit, the MTF becomes nonlinear and actually better than the above approximation. Beyond a ratio of 0.7, the divergence is large. For a slightly better accuracy, substitute 1.27 for the 1.3 in the first equation.

In the LWIR, one can assume a 2 cycle/mrad for a one-inch aperture, and a 5 cycle/mrad in the MWIR. Thus, a derivative rule of thumb is that the spatial frequency cutoff is 2.5 times the aperture in inches (one-half due to Nyquist) for the MWIR, and twice that for LWIR.

The modulation transfer function of a lens will be somewhat less than the perfect case considered above. Nevertheless, when rigorously derived, it is almost linear with spatial frequency (resolution) and can be approximated by the above relationship, but cautious individuals would degrade it slightly.

Sources

G. Waldman and J. Wooton. 1993. *Electro-Optical Systems Performance Modeling.* Norwood, MA: Artech House, 125.

L. Levi. 1968. *Applied Optics,* vol. 1, *A Guide to Optical System Design.* New York: John Wiley & Sons, 487–488.

Additional information supplied by W. M. Bloomquist, 1995.

Radiometry

Radiometry is the study of the measurement and radiative transport of electromagnetic radiation, independent of wavelength; often the term is also used to include the detection and determination of the quantity, quality, and effects of such radiation. Conversely, *photometry* describes this for the visible portion of the spectrum only. They are the same concept, except photometry and its inane terms and dimensions are a result of normalizing (or attempting to normalize) the measurement of light to the response of the human eye.

William Herschel (1738–1822) not only discovered infrared radiation, but he attempted to draw the first distribution of energy as a function of wavelength. Johann Lambert (1728–1777) noted that the amount of radiated (and in some cases reflected) energy in a solid angle is proportional to the cosine of angle between the emitter and receiver. (Lambert also proved that π is an irrational number and introduced the hyperbolic functions sinh and cosh.) A few decades later, Gustav Kirchoff (1824–1887) discovered that the emissivity of a surface is equal to its absorption. Later, Austrian physicist Josef Stefan (1835–1893) determined the total radiant exitance from a source from all wavelengths to be equal to the emissivity multiplied by a constant (the Stefan-Boltzmann constant) times its temperature raised to the fourth power. In 1866, Langley used a crude bolometer to study the radiation of carbon at different temperatures. In 1886, Michelson employed Maxwell's laws to develop crude blackbody laws. Additionally, two famous rules of thumb (or useful approximations) that led to the Planck function were described by Wein and Rayleigh. Jeans found a numerical error in Rayleigh's equation, so it is now known as the Rayleigh-Jeans law. Others, such as Lummer and Pringsheim, made important pre-Planckian additions to blackbody theory.

However, the main architect of modern radiometry was Max Karl Ernst Ludwig Planck (1858–1947). Max Planck began his scientific career under the influence of Rudolf Clausius (developer of the second law of thermodynamics), giving him a strong background in thermal physics. Planck is most noted for describing the blackbody radiation in a simple equation and for developing the quantum theory of energy (which states that energy is not infinitely subdividable but exists in quantum pieces). He noted that the Rayleigh-Jeans approximation agreed well with experimentation for long wavelengths and that the Wein law worked well at shorter wavelengths, and he devised a mathematical approximation that would incorporate both at the proper wavelengths. On October 19, 1900, he presented this, and within a few days experiments were constructed that supported his new equations. William Coblentz (1873–1962) was the first to experimentally verify Planck's law by accurately measuring blackbody radiation. He continued to refine his equations until, on December 14, 1900, he presented the law as we now know it. He was given the 1918 Nobel prize for his work. Interestingly, Planck remained skeptical of his theory until his death. Although there is some speculative evidence that Planck's blackbody equation is not absolutely correct to an infinite number of decimal places, recent evidence leads some researchers to postulate that Planck's theories are analogous to Newton's laws of mechanics. Newton's mechanics are correct under certain conditions (the ones usually present to mortals) but needs the refinement of relativity to be applicable to other conditions. The task of refining Planck's laws is left to future generations (with more accurate experimental instrumentation) as an exercise.

Often not appreciated today, describing blackbody radiation represents a momentous achievement in science and engineering. Planck's spark of creativity in radiometry allowed succeeding scientists to explore nature under a new paradigm. Planck's theories led to the explanation of the photoelectric effect (for which Albert Einstein won the Nobel prize), the Bohr model of the atom (the springboard for all modern particle physics), quantum mechanics, and quantum electrodynamics, and provided additional verification of Maxwell's equations.

Although instrumental, Planck was not alone in the development of the foundations of radiometry. Many, like Newton, Melloni, Bouguer, Leslie, and others also contributed greatly. The blackbody function was crudely measured by observing glass melts. This experimental data helped the early theorists form the foundation for the various early theories of blackbody radiation.

Planck described the blackbody radiation as an oddly shaped curve based on the temperature raised to the fourth power and the "quantum action" of the atom. To calculate the energy within a given spectral bandpass, the curve must be integrated with the spectral cut-on and cutoff as the limits on the integral. Most mortals find this difficult to do mentally. So, before every engineer had a portable computer and a calculator, there existed a plethora of rules of thumb and nomographs to approximate the Planck function. In fact, almost every book published before about 1990 on electro-optics has a chapter or two dedicated to the Planck function and the authors' favorite short cuts. Many are elegant, but some others are no longer necessary. A few are included in this chapter to help develop a feel for the blackbody function and to use when it is inconvenient to whip out the portable computer.

Recent advancements in the study of radiometry include the many incremental improvements by William Coblentz and Fred Nicodemus. Even as we approach the year 2000, radiometric measurements are plagued by the immaturity of instruments and subtle variations in experimental apparatus, making highly accurate measurements difficult. It is an arduous (and sometimes impossible) task to repeatedly and accurately determine the effects of equipment instability, specular radiation, unwanted reflections, and the atmosphere. With the exception of some narrow lines, suitable standards for much of the spectrum are in their infancy, with stability, measurement repeatability, and associated transfer standards limited in accuracy to a few percentage points. As a result, most general measurements are accurate only to within a few percentage points of theoretical calculations. Further advancements in this field are likely to begin with more accurate measurement instruments and techniques, and increased accuracy of transfer standards.

For those interested in more detailed information on radiometry, the authors suggest Volume 1 of *The Infrared and Electro-Optical Systems Handbook,* and appropriate chapters in the following books:

- *The Infrared Handbook* (Wolfe and Zissis)
- *Electro-Optics Handbook* (Burle)
- *Infrared System Engineering* (Hudson)
- *Selected Papers on Infrared Design* (Johnson and Wolfe)
- *Far Infrared Techniques* (Kimmitt)
- *Electro Optical Imaging System Performance* (Holst)
- *Electro-Optical Systems Performance Modeling* (Waldman and Wootton)
- *Radiometric Calibration* (Wyatt)
- *Electro Optical Systems Analysis* (Seyrafi)

Journals that often contain papers on radiometry include: *Optical Engineering* and *Infrared Physics and Technology.* Finally, do not forget to monitor the NIST home page on the World Wide Web (http://physics.nist.gov).

BLACKBODY OR PLANCK FUNCTION

Subject: Radiometry, target phenomenology

The Rule

Radiant exitance as a function of wavelength is:

$$M_{\lambda\,(\mu m)} \le \frac{c_1}{\lambda^5 \left(e^{c_2/\lambda T} - 1 \right)} \left[\frac{\text{watts}}{\text{area}} \right]$$

where M_λ = radiant exitance in watts per unit area per 2π sr per micron at λ
   π = pi (3.14159 . . . .)
   $c_1 = 2\pi c^2 h$, sometimes called the *first radiation constant*, = 37418 W $\mu m^4/cm^2$
   λ = wavelength in microns (or in alternative units to match c_1 and c_2)
   c_2 = the second radiation constant, hc/k, where k is Boltzmann's constant; c_2
      is equal to 14388 μm K for the above units (or 1.4388 cm K)
   T = temperature in Kelvins

Likewise, this can be written for a photon flux with a slightly different form.

$$M_q \le \frac{2\pi c}{\lambda^4 \left(e^{c_2/\lambda T} - 1 \right)} \left[\frac{\text{photons}}{\text{sec area}} \right]$$

where c = speed of light in a vacuum
   h = Planck's constant

Basis for the Rule

This was developed by Planck in 1900. He used an ingenious combination of empirical (e.g., data from Langley) evidence and theoretical statistics to explain Wein's formulas and the Rayleigh-Jeans rule. Planck hypothesized that energy is not emitted continuously but in discrete quanta (hc/λ), the energy absorbed or emitted depending on the wavelength, and that radiation arises from linear vibrations of atoms analogous to Hertz oscillators.

Cautions and Useful Range of the Rule

This rule is for a blackbody with an emissivity of 1. For lower emissivities, this should be multiplied by the lower emissivity.

   Some real-world items are spectral emitters and do not follow a blackbody curve over broad bandpasses. Examples of spectral emitters include H_2O and CO_2 in rocket plumes.

   When using this rule, be sure that all of the units match; using these equations with mismatched constants and units is a frequent error.

   The radiance (or *sterance*) of a blackbody is $1/\pi$ times the above radiant exitance.

Usefulness of the Rule

This rule employs a fundamental equation that governs the nature of thermal emission and is used throughout the photonic disciplines. It is used for determining the blackbody emittance or graybody emittance (by multiplying the above by the lower emissivity).

The in-band radiant exitance depends on the wavelength limits, short (s) and long (l), and can be expressed as

$$\int_{\lambda_s}^{\lambda_l} M_\lambda\, d\lambda \text{ for watts or } \int_{\lambda_s}^{\lambda_l} M_q\, d\lambda \text{ for photons/second}$$

can be easily calculated using the above equations and a spreadsheet via numerical integration. To do so, simply:

1. Program a spreadsheet with the above equation (your choice of M_λ or M_q).

2. Multiply the equation by an incremental wavelength (say 1/100th of your bandpass).

3. Set λ_l equal to your short-wave cut-on.

4. Increment λ_s by your chosen increment.

5. Repeat for the number of increments that you have (calculate it 100 times).

6. Add up the numbers and divide the final result by the number of increments.

Notes and Explanation

Sometimes $2\pi c$ is called the *third radiation constant* (c_3) and is equal to 1.884×10^{23} $\mu m^3/s\ cm^2$.

The $\leq$ is used in the above equations to indicate that real-world objects are graybodies with an emissivity of less than 1, so their radiant output is always somewhat less than what would be predicted for a perfect blackbody.

It is interesting to realize that the average number, n, of photons per degree of freedom is emitted by an object at a temperature, T, is

$$\frac{1}{\exp\left(\dfrac{h\upsilon}{kT}\right) - 1} = \frac{1}{\exp\left(\dfrac{c_2}{\lambda T}\right) - 1}$$

and, hence, it appears in the above equations.

When the first equation is integrated from zero to infinity, the Stefan-Boltzmann law results, which give the total radiant exitance (σT^4).

Planck's original development of this gave birth to quantum theory.

PLANCK FUNCTION APPROXIMATION WITH WAVENUMBER

Subject: Radiometry, spectrometers

The Rule

In terms of wavenumbers, the Planck function can be approximated and simplified to

$$F_\upsilon = 8.3 \times 10^{-13} \upsilon^2 d\upsilon \quad \text{in W/cm}^2/\text{sr/wavenumber/K}$$

where F_ν = flux of radiation per unit source area per solid angle per wavenumber per Kelvin
ν = wavenumber (defined as $1/\lambda$)

Basis for the Rule

For high temperatures and long wavelengths, $h\nu c$ is much less than KT, so the above can be written. This rule is based on the Rayleigh-Jeans law, which is a simplification of the Planck function using wavenumbers. Consider the following:

$$M_\lambda \Delta\lambda_{(cm)} \approx \frac{2\pi ckT}{\lambda^4} \Delta\lambda_{(cm)}$$

Converting to wavenumber yields

$$\upsilon = \frac{1}{\lambda} m^{-1}$$

$$\Delta\upsilon = -\frac{\Delta\lambda}{\lambda^2}, \text{ so } \Delta\lambda = -\lambda^2 \Delta\upsilon$$

$$M_\lambda \Delta\lambda = 2\pi ckT\upsilon^4 (-\lambda^2) \Delta\upsilon = -2\pi ckT\upsilon^2 \Delta\upsilon$$

The minus sign indicates that an increase in $\Delta\lambda$ produces a decrease in $\Delta\nu$.

Cautions and Useful Range of the Rule

This rule is valid when the product of the wavelength and temperature is greater than 10^{15} μm K (or when $T/\upsilon > 10$ K/wavenumber). Thus, this rule is valid as an approximation for long wavelengths and hot temperatures. It is not perfect, but it is a good approximation.

Usefulness of the Rule

This rule is useful for rapid estimations of the signal per wavenumber from an object of a given temperature.

Notes and Explanation

Wavenumbers are properly defined as $1/\lambda$. One of the benefits of thinking in wavenumbers is that the Planck function is almost linear for long wavelengths (say, >10 or 15 μm) and hot temperatures (say, >350 K).

Source

M.F. Kimmitt. 1970. *Far-Infrared Techniques*. London: Pion, 43–45.

THE RAYLEIGH-JEANS APPROXIMATION

Subject: Radiometry, estimating spectral radiance

The Rule

Rayleigh and Jeans developed the following equation to describe the emission of spectral radiance:

$$L_\lambda \approx \frac{2\pi k c T}{\lambda^4}$$

where L_λ = spectral radiance in W m^{-2} sr^{-1}
k = Boltzmann's constant = 1.38×10^{-23} J/K
T = temperature in Kelvins
λ = wavelength of the light
c = speed of light = 3×10^8 m/s

This same approximation holds in the wavenumber (almost akin to frequency) domain and takes on the form:

$$L_\kappa \approx \frac{2 k T \kappa^2}{c^2}$$

where L_κ = spectral radiance in W m^{-2} wavenumber^{-1}
κ = wavenumber, defined as $1/\lambda$
c = speed of light = 3×10^8 m/s

Basis for the Rule

This represents a mathematical approximation of Planck's law for long wavelengths and high temperatures relative to the peak wavelength. The above relationships were derived from classical statistical mechanics.

Cautions and Useful Range of the Rule

The value of $h\nu$ must be much smaller than that of KT, which implies long wavelengths relative to the peak. This also assumes Lambertian blackbodies.

Usefulness of the Rule

This approximation can be used for quick estimates of signature in a narrow band.

Notes and Explanation

When John Rayleigh (1842–1919) was not working on scattering, he suggested the above approximation, which was separately refined by James Jeans (1877–1946). This provides a very simple way of calculating radiant exitance in the more linear portion of the Planck function. Spectroscopists will find the approximation more useful when expressed in frequency (the second equation). This was called the "UV catastrophe" by Planck and others, as it yields increasing energy with shorter wavelengths without limit. Clearly, this was not the case, and Planck proved it so at the turn of the century with his complete expression, which established the basis of quantum mechanics:

$$Q_p(T) = \int_{\lambda_s}^{\lambda_l} \frac{2\pi c}{\lambda^4 [e^{(hc/\lambda kT)} - 1]} d\lambda$$

$$W(T) = \int_{\lambda_s}^{\lambda_l} \frac{2\pi c^2 h}{\lambda^5 [e^{(hc/\lambda kT)} - 1]} d\lambda$$

where $Q_p(T)$ = photon flux in photons/s/cm^2/sr in the band defined by λ_s and λ_e
  λ_l = long wavelength band cutoff
  λ_s = short wavelength band cut-on
  c = speed of light
  λ = wavelength
  h = Planck's constant
  k = Boltzmann's constant
  T = temperature
 $W(T)$ = radiant flux per unit area in W/cm^2

NARROWBAND APPROXIMATION OF PLANCK'S LAW

Subject: Radiometry, Planck's law, detectors

The Rule

Martin[1] gives a narrowband approximation of the Planck radiation law as

$$\Phi_\lambda = \frac{2c\Delta\lambda}{\lambda^4}e^{-1.44/\lambda T}$$

where Φ_λ = flux in photons per square cm per second per steradian at the center
   wavelength of the band defined by $\Delta\lambda$
   c = speed of light (3×10^{10} cm/s)
   $\Delta\lambda$ = difference in wavelength across the bandpass (in cm) $(\lambda_l - \lambda_s)$
   λ = median wavelength (in cm)
   T = temperature of interest (e.g., the background or scene temperature) in
   Kelvins

Basis for the Rule

Planck's blackbody integral law becomes algebraic when the difference between the
upper and lower wavelengths' difference is less than about half a micron. This
closed-form expression of the Planck function (with units of photons per second
per square centimeter per steradian) varies exponentially as:

$$\Phi_\lambda(T) = \Phi_\lambda e^{-X(\lambda)/T}$$

where $\Phi_\lambda(T)$ = photon flux at a given wavelength for a given blackbody temperature
   $\Phi_\lambda = 2c(\Delta\lambda)\lambda^{-4}$
   $X(\lambda) = hc/k\lambda = 1.44/\lambda$ when λ is in units of cm
   T = temperature in Kelvins

Cautions and Useful Range of the Rule

This rule works well for hyperspectral and multispectral systems where each bandpass is less than a quarter of a micron. The narrower the strip of the spectrum, the
more accurate this approximation becomes. Typically, the accuracy is the ratio of
the width of the strip to the wavelength.

   This rule assumes a narrow band, for bands of one micron or less.

   This rule assumes an item of interest (e.g., scene or target) of a constant temperature and emissivity.

   Generally, this provides an accuracy to within a few percentage points for bands
less than one micron in width.

Usefulness of the Rule

This is useful for a narrowband approximation of Plank's radiation law, when calculate in photons. This is very useful in normal IR practices when attempting to calculated the amount of background flux on a detector (which may limit integration
time by filling up the wells), with the square root being the associated noise.

For wide bands, the photon flux can be found by adding up successive pieces of the bandpass using the above equation (e.g., numerical integration on a spreadsheet).

This narrow band approximation is useful because its derivative at a specified temperature can be taken in closed form to allow the contrast function to be found for a given temperature.

Notes and Explanation

Planck's blackbody function can be reduced to this narrow approximation when applied to a given wavelength and a target at a constant temperature.

Aficionados will realize that the 1.44 relates to the classic radiation constant derived from (hc/k)

Example

Consider the solar spectrum and assume the sun is a 5770 K blackbody source (see the rule "Blackbody Temperature of The Sun," page 3), and we will calculate the photon flux for a 1μm wide bandpass at 0.6μm. Then,

$$\Phi_\lambda = \frac{2c\Delta\lambda}{\lambda^4}e^{-1.44/\lambda T} = 4.63\times10^{27}\ e^{-4.16} = 7.23\times10^{25}\ \text{photons/s cm}^2\Delta\lambda\,(\text{cm})$$

If we multiply 7.23×10^{25} by a 1 micron bandpass (10^{-4} cm), we get 7.2×10^{21} photons/s cm^2. Using standard theory, the flux is 6.1×10^{21} photons/s cm^2, so the rule provides reasonably accurate results.

Reference

1. Rule provided by Dr. Robert Martin, 1995.

LOGARITHMIC BLACKBODY FUNCTION

Subject: Radiometry, blackbody emission

The Rules

When we plot a blackbody's output (W/m²/sr) wavelength on log-log graph paper (see Fig. 14.1), the following is true:

1. The shape of the blackbody radiation curve is exactly the same for any temperature.

2. A line connecting the peak radiation for each temperature is a straight line.

3. The shape of the curve can be shifted along the straight line connecting the peaks to obtain the curve at any temperature.

Basis for the Rules

The rules are simply based on the Planck function.

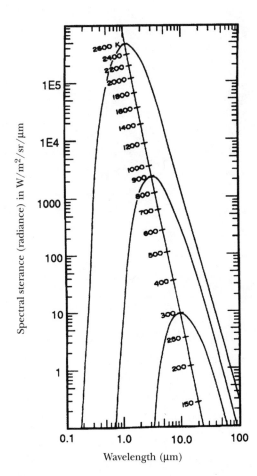

Figure 14.1 Spectral Sterance vs. Wavelength and Temperature (from Ref. 1, courtesy of Academic Press)

Cautions and Useful Range of the Rule

For the rules to be true, we must use a log-log plot. It is valid for blackbody radiation only (e.g., not spectral emitters).

Usefulness of the Rule

The rules allow for a simple determination of the blackbody function.

Notes and Explanation

Spectral irradiance can be determined quickly from Fig. 14.1. First, trace the general shape of the Planck function. Place it on the graph, matching the peak of your trace to the tilted peak line. The spectral irradiance can be determined by moving the trace up and down that line, setting the peak to the desired temperature.

Reference

1. C. Wyatt. 1978. *Radiometric Calibration, Theory and Methods* Orlando, FL: Academic Press, 32–33.

THE DIFFERENTIAL RADIANCE RULE

Subject: Radiometry, systems, figures of merit

The Rule

Hudson[1] estimated the differential radiance for a 300 K object within three popular atmospheric bands as follows:

SWIR	2.0 to 2.5 μm	6×10^{-9} W/cm^2/sr/K
MWIR	3.2 to 4.8 μm	5.2×10^{-9} W/cm^2/sr/K
LWIR	8 to 13 μm	7.4×10^{-5} W/cm^2/sr/K

Basis for the Rule

This rule is based on the differentiation of the Planck function integrated over the stated bandpasses.

Cautions and Useful Range of the Rule

This concept is very bandpass sensitive; a slight difference between the above bands and the one that you are using will reduce the accuracy.

The above are calculated for a departure from 300 K, so if your target is more than a few degrees away from that level, then accuracy is degraded. You might want to calculate the differential radiance once for your bandpass and target temperature and keep it handy for future calculations.

Assuming a flat response system, this is valid over the stated bandpass.

Usefulness of the Rule

The rule is useful for quick conversion between NEΔT and *noise equivalent irradiance* (NEI) (see example). This rule can also be applied for a quick estimation of the radiance effects from a change in temperature, for NEΔT calculations, and to underscore the amount of radiance produced by an object.

Notes and Explanation

Differential radiance is the radiance per an increment of temperature (1° C or K). It is important in NEΔT calculations.

Frequently, one needs to calculate a NEΔT or a radiance from a given NEΔT. This involves an integral of the Planck function with the bandpass as the limits of the integral. For the atmospheric bands, one can easily estimate this from the above.

Note that a 1 K change in temperature from a 300 K object produces a change that is higher in the LWIR than MWIR by several orders of magnitude. However, the contrast difference for typical thermal objects (say, near 300 K) is still greater in the MWIR than the LWIR. This is because the ratio (or percentage) of the differential in-band radiance to the total in-band radiance is greater for MWIR.

Example

If a LWIR system has a stated sensitivity of 0.01° C NEΔT and a 1 mrad field of view, we can easily estimate the NEI by multiplying the above conversion factor by the NEΔT and the field of view in steradians (area or square FOV) as follows:

$$(7.4{\times}10^{-5})\,(0.01)\,(0.001)^2 = 7.4{\times}10^{-13}\ \text{W/cm}^2$$

Reference

1. R. Hudson. 1969. *Infrared Systems Engineering*. New York: John Wiley & Sons, 431.

LAMBERT'S LAW

Subject: Radiometry, phenomenology, targets

The Rule

Most surfaces can be considered to be diffuse (or "Lambertian"), and their emittance and reflection are as follows:

$$L = M/\pi$$

where L = radiance in watts/m^2/sr
M = radiant exitance in W/m^2

or, more specifically and in terms of radiant intensity,

$$I_\theta \propto I\cos\theta$$

where I_θ = radiant intensity in W/sr from a surface viewed at an angle θ from the normal to the surface
I = emitted radiant intensity in W/sr
θ = viewing angle between the normal to the emitting surface and the receiver

Basis for the Rule

The rule is based on theory and empirical observation as originally described by Lambert.

Cautions and Useful Range of the Rule

The rule assumes that the surface is not specular, which is a good assumption unless you know otherwise.

As wavelength increases, a given surface is less likely to be Lambertian.

As the angle of incidence decreases, the surface is less likely to exhibit Lambertian properties. At grazing incidences, most surfaces exhibit a specular quality.

A surface actually radiates and received energy from 2π steradians, as first would be expected. The decrease to π is due to the projected angle effects (cosine) and is the resultant value to use in calculations.

Usefulness of the Rule

This rule is useful in estimating reflectances and emissions from surfaces and is indispensable for radiometric calculations.

Notes and Explanation

Unless special preparation is take (such as polishing), most surfaces reflect and emit as a diffuse surface. This is because most surfaces are "rough" at the scale of the wavelength of light. Therefore, they reflect and emit their radiation following Lambert's cosine law and are called "Lambertian." Mirrors do not follow this law but, rather, the laws of reflection for geometrical optics.

Although simple, the first equation represents a powerful rule. It enables one to quickly change from a spectral emittance defined by the Planck function to a spectral radiance by merely dividing by π. This works for reflection as well. Note that it is

M/π, not M over 2π. The conversion factor is just one π. This results from the phenomenon of a projected solid angle (the surface actually does emit photons into 2π). The projection of the hemisphere (2π) in which the object is radiating into onto a two-dimensional flat surface yields a disk (π). Thus, their reflection and emission as a function of angle is merely the emission divided by π.

Additionally, when a sensor is viewing a surface at an angle, the radiant intensity is decreased by the cosine of the angle between the surface emitting (target) and the surface receiving (aperture). This explains why a spherical illuminated object (e.g., the sun) appears to have equal brightness across its surface. If the radiant intensity did not decrease as the cosine, but was constant, then the edges would appear brighter. This is because you observe more surface area at the edges due to the curving. However, in reality the extra area observed is exactly compensated for by the above cosine effect. This is why the sun appears as a disk of constant brightness.

In the second equation, the radiant intensity is stated as a direct proportion rather than an equality. This because the media (e.g., atmosphere) between the emitter and receiver may alter the perceived radiant intensity as well as an emissivity of less than unity. Incidentally, radiant intensity can be changed to the more useful irradiance (flux) merely by dividing the range squared.

ABSORBED, TRANSMITTED, OR REFLECTED

Subject: Radiometry, physics

The Rule

When light strikes a surface, it is absorbed, transmitted, and reflected. A little of all three usually occurs, but it all goes somewhere.

Basis for the Rule

This is based on the principle of conservation of energy.

Cautions and Useful Range of the Rule

It is not always apparent how much is being absorbed, transmitted, and reflected. The distribution in one part of the spectrum does not necessarily reflect (pun intended) what is happening in other parts of the spectrum.

Usefulness of the Rule

This is useful as a reminder that all photons must be accounted for, and that they are always distributed throughout the various interaction processes.

Notes and Explanation

If a single photon strikes a surface (such as your retina or the front surface of the first optical element of a telescope), there are three possible outcomes. First, it may be absorbed and transformed into thermal energy, thereby raising the temperature of the object. This tends to occur in dielectric materials and semiconductors (e.g., glass and detectors). The second possibility is that it will be transmitted through, which is what usually happens with dielectrics and semiconductors when the photon's energy is less than what is needed to free an electron. The third option is that it will be reflected or scattered, which usually happens for a metal. Unfortunately, a lot of it will be scattered to places where you do not want it.

However, a light beam is not made of a single photon, but a whole slew of them. All of these processes happen all of the time, so the beam is divided into reflection, absorption, and transmission. The mixture of the three processes varies greatly, depending upon material, thickness, and angle of incidence.

The Peak Wavelength, or Wein Displacement, Law

Subject: Radiometry, blackbody radiation curves

The Rule

The peak wavelength (in microns) of a blackbody is approximately 3000 divided by the temperature in Kelvins.

Basis for the Rule

This is an approximation and generalization of blackbody theory.

Cautions and Useful Range of the Rule

This assumes the emitter is a blackbody or graybody and not a spectral emitter. Thus, it does not take into account potential emissivity changes as a function of wavelength. However, unless such changes are very severe, this is not likely to change the location of the peak wavelength.

Usefulness of the Rule

This little rule provides a quick determination of peak wavelength as well as a basis to quickly estimate the energy distribution.

Notes and Explanation

According to Planck's law, a blackbody will have an energy distribution with a unique peak in wavelength. For a blackbody, this peak is solely determined by the temperature and is equal to $2898/T$. Hudson[1] points out that about 25 percent of the total energy lies at wavelengths shorter than the peak, and about 75 percent of the energy lies at wavelengths longer than the peak

Additionally, Hudson gives the following short cuts:

> To calculate the wavelengths where the energy is half of the peak (half power or at the 3 dB points), divide 1780 by the temperature in Kelvins for the lower end and 5270 for the higher end. You will then find:

- 4 percent of the energy lies at wavelengths shorter than the first half power point
- 67 percent of the energy lies between the half points
- 29 percent of the energy lies at wavelengths longer than the longest half power point

Reference

1. R. Hudson. 1969. *Infrared Systems Engineering.* New York: John Wiley & Sons, 58–59.

PHOTONS-TO-WATTS CONVERSION

Subject: Radiometry, unit conversion

The Rule

To covert a radiometric signal from watts to photons per second, multiply the number of watts by the wavelength (in microns) and by 5×10^{18}.

$$\text{Photons/s} = (\lambda_{[\text{in microns}]}) \; (\# \text{ of watts}) \; (5 \times 10^{18})$$

Basis for the Rule

This rule is based on the definition of a watt coupled with basic nature and physics.

Cautions and Useful Range of the Rule

Actual results are only for a given wavelength or an infinitesimally small bandpass. Typically, using the center wavelength is accurate enough for lasers or bandwidths of less than 0.2 μm. However, if doing this conversion for wide bandpasses, set up a spreadsheet and do it in increments of, say, $1/20$ μm.

This is only valid if the wavelength is expressed in microns. The constant must be adjusted for wavelengths expressed in other units.

Usefulness of the Rule

It can be used for converting watts to the number of photons per second or vice versa.

Notes and Explanation

If you require more than two significant figures, the constant is 5.0345×10^{18}.

The actual conversion can be stated as watts $= (hc/\lambda)$ photons/s, so photons per second $= (\text{watts} * \lambda / hc) = \text{watts} \times \lambda$ (in meters) $\times (5 \times 10^{24})$ for all terms with the dimension of meters (wavelength in meters). There are a million microns in a meter, so the constant is 10^6 smaller if one uses microns.

NEDT SIMPLIFICATION

Subject: Radiometry, systems, figures of merit

The Rule

$$NEDT^*_{\text{ideal}} (\lambda_{co}) \approx KT^2 D^*_{\text{ideal @ } \lambda co}$$

where $NEDT^*_{\text{ideal}}$ = ideal noise equivalent delta temperature achievable from a 300 K nominal for a given spectral cutoff

k = Stefan-Boltzmann's constant ($1.38{\times}10^{-23}$ Joule/K)

$D^*_{\text{ideal @ } \lambda co}$ = specific detectivity of an ideal photoconductor with a cutoff wavelength of λ_{co}

T = temperature in Kelvins

In addition,

$$NEDT^*_{\text{min}} = 5.07{\times}10^{-8} \sqrt{\frac{300}{T}} \text{ (K cm s}^{1/2})$$

where $NEDT^*_{\text{min}}$ = minimum noise equivalent temperature difference achievable for a 300 K background when the background is the limiting factor in performance (called the BLIP) condition

Basis for the Rule

These rules are based on definitions of NEDT and basic radiometric principles applied to a condition when the shot noise of the arrival of the photons dominates (a BLIP condition).

Cautions and Useful Range of the Rule

BLIP conditions are assumed. In general, to a first approximation, the $NEDT^*_{\text{ideal}}$ can be approximately scaled by

$$\frac{D^*_{\text{ideal}}}{D^*_{\text{actual}}}$$

However, as always when using D*, one must beware of the D* measurement parameters (does it include $1/f$ noise, readout noise, cold shield, and so on?) and be conscious of well size.

The second equation assumes a 300 K nominal background from which the difference is derived.

The rule does not account for the bandpass cut-on, although Taubkin et al.[1] explain that this is generally of little concern for the theoretical minimum.

As always, NETD implies an imaging system (not applicable to point source detection systems).

The user should consider these as approximations.

Usefulness of the Rule

As photodetectors get better and better, BLIP conditions are more often achieved. The above rules give a simple approximation for the minimum (best) NEDT achievable. Anyone claiming better system performance is blowing smoke. Although the authors cannot think of a single system that achieves the incredibly small NEDT mentioned above, such systems are probably just beyond your current calendar.

Notes and Explanation

For a BLIP case, the sensor hardware noise is negligible compared to the photon flux variations. Therefore, the minimum NEDT can be calculated depending on temperature only. For the popular 3- to 5-μm band, this works out to be 2.7×10^{-7} K, and for the 8 to 12 band popular in America, the minimum is 8×10^{-8}. For the 8 to 14-μm band, which is popular in Europe, it is 7×10^{-8}. However, the shorter bandpasses tend to compensate for this difference by having higher contrast.

Reference

1. I. Taubkin et al. 1994. Minimum temperature difference detected by the thermal radiation of objects. *Infrared Physics and Technology* 35(5), 718.

THE MRT/NEDT RELATIONSHIP[1]

Subject: Radiometry, systems, figures of merit, ergonomics

The Rule

For thermal imaging sensors, the sensitivity of a system varies inversely to the system MTF.

For IR systems, the *minimum resolvable temperature* (MRT) can be forecast from the *noise equivalent delta temperature* (NEDT) by:

$$MRT \approx k\,(NEDT)\,/\,(MTF_{sys(f)})$$

where k = proportionality constant (Holst indicates that this should be between 0.2 to 0.5)

$MTF_{sys(f)}$ = system-level modulation transfer function at spatial frequency f

Basis for the Rule

This is based on empirical observations and human interactions with FLIRs.

Cautions and Useful Range of the Rule

This is an approximation only, accurate for mid-spatial frequencies, and it applies to thermal imaging IR sensors only.

Operator head movement is allowed in this relationship

MTF and MRT must be at the same spatial frequency, and MTF should account for LOS stability and stabilization.

Usefulness of the Rule

An awareness of this relationship is useful because it underscores the effect or MTF on system performance.

It can be used to estimate a hard-to-calculate quantity (MRT) from two easily calculable (or determined) quantities (NEDT and MTF).

It may be used to estimate the MRT when an MRT test (or its data) is not convenient (or available).

It illustrates (and reminds) us of some of the drivers on MRT.

Notes and Explanation

Sensitivity of thermal imaging sensor systems seems to fall off linearly with a decrease in the MTF. For IR systems, this can be related to the NEDT, and the MRT is usually somewhere between 1/2 and 1/5 times the NEDT.

Reference

1. G. Holst. 1989. Minimum resolvable temperature predictions, test methodology and data analysis. *Proc. SPIE*, vol. 1157, 208–216.

QUICK TEST OF NEDT

Subject: Radiometry, systems, medical instrumentation

The Rule

If an infrared system can image the veins in a person's arm, then the system is achieving a NEDT or MRT (whichever is proper) of 0.2 K or better.

Basis for the Rule

This is based on radiometry and empirical observations.

Cautions and Useful Range of the Rule

This provides crude approximations only. It does not account for adverse effects such as focusing, atmospheric transmission, and the like.

It does not tell you how good the system is for subpixel detection—just that it is better than a NEDT of 0.2 K.

It seems to be more difficult to image the vein in a woman's arm than a man's. This is probably due to the extra layer of fat and less muscle. Imaging a woman's arm veins would often indicate an even better NEDT.

Usefulness of the Rule

This is useful for quick estimates of system performance. This is especially useful when the camera is not yours and is publicly set up at a conference, lab demo, or some other function.

Notes and Explanation

Because they are transporting warm blood, the veins in a person's arm or head tend to have a temperature difference of 0.1 to 0.3° C above the outside skin temperature. If your camera can image them, then it has a NEDT of this amount or better (smaller).

NOTHING IS PERFECT, OR THERE ARE NO PEDIGREE PAPERS IN RADIOMETRY

Subject: Radiometry, environment

The Rule

Nothing is a blackbody. Nothing is a graybody. Nothing is specular. Nothing is Lambertian.

Basis for the Rule

This is based on the fact that, while we consider things to be "black" or Lambertian, nothing is perfect. A black surface will always have some small amount of reflection. This is also derived from observations of the world and definitions of terms.

Cautions and Useful Range of the Rule

In some situations, some objects approach pure blackbodies or pure specular surfaces to levels that your sensors or eye cannot dispute.

Usefulness of the Rule

Keeping this rule in mind underscores the nature of the world and instills caution in the system designer and system tester.

Notes and Explanation

We, our instrumentation, backgrounds, and targets exist in a world that is not as perfect as simple mathematics would dictate. Everything is a combination of a specular emitter and a blackbody to different degrees. Likewise, everything is a combination of a diffuse Lambertian surface and a specular mirror surface. This is not to indicate that these concepts are not valid; *they are*. This is not to indicate that an object cannot exhibit one of these characteristics predominantly; *it can*. This is to say that, when modeling the real world, allow for some slack to represent reality.

CALIBRATE UNDER USER CONDITIONS

Subject: Radiometry, calibration, systems

The Rule

The calibration of an instrument for a specific measurement should be conducted in a way that makes the resulting measurement independent of instrument effects. Moreover, the calibration should be conducted under conditions that reproduce the situations under which the field measurements are to be made.

Basis for the Rule

This rule is based on radiometric principles and the way sensor systems behave.

Cautions and Useful Range of the Rule

Sometimes calibrations of sufficient accuracy may be done in slightly different environments. The radiometric and environmental conditions should simulate the range of conditions that are likely to affect the results; they do not need to exceed those conditions.

Usefulness of the Rule

This rule is useful when designing calibration facilities, determining calibration requirements, and understanding the usefulness of a previous calibration.

Notes and Explanation

The purpose of a calibration is to establish a known relationship between the output of an electro-optical sensor and a known level of incident flux that can be traced to primary standards (e.g., NIST in the U.S.A.). As stated above, when the flux is to be measured at different times and places, or with different instruments, the results should be the same (or of a known function of each other). Sensor contributions should not be measured, or measured and understood sufficiently to allow their subtraction.

Additionally, electro-optical sensors have quirky attributes that make their output a complex, nonintuitive function of the total input. When attempting to calibrate a sensor, the calibration test should be conducted in an environment as close as possible to that expected in the sensor's use. This includes the spectral radiant flux from a background, the temperature of the sensor, the expected radiant flux on the aperture, similar target geometry, background, polarization, vibration, and so on. Also, measurements must be made a sufficient number of times to determine repeatability and an estimate of calibration uncertainty.

ABSOLUTE CALIBRATION ACCURACY

Subject: Radiometry and calibration

The Rule

Typically, absolute calibration to a national standard cannot be done to a traceable accuracy better than the percentages shown below.

Longwave IR	5–10 percent
Shortwave IR	2–5 percent
Near infrared	2 percent
Visible	1 percent
Ultraviolet	1 percent

Basis for the Rule

Errors add up. By the time you calibrate your instrument, the transfer errors (e.g., variations in repeatability, stray light, unwanted reflections, variations in emissivity, variations in temperature, temperature measurement accuracy, and the like) will limit your absolute calibration.

Cautions and Useful Range of the Rule

This is based on the state of the art of transfer standards, calibration sources, and facilities. These are constantly being improved. Better accuracy is theoretically possible with additional development of sources and test facilities.

Sometimes, better accuracy can be achieved for an extremely narrow bandpass that happens to have a easily transferred source (especially in the UV and visible). This rule relates to absolute calibration; relative calibration for short periods can be an order of magnitude better.

Usefulness of the Rule

This is useful for determining calibration requirements, specifications, and real-world performance. When desperate, use the above as a baseline for inputs to algorithms and system studies.

Notes and Explanation

Sometimes this rule is stated as the *Melloni Haunts Your Measurements Rule*. Macedonio Melloni (1798–1854) was an early researcher of radiometry. It could be arguable said that Melloni was the founder of electro-optics, as he and Leopoldo Nobili constructed the first electro-optical detector (an antimony and bismuth thermopile). Melloni was an experimental genius, well ahead of the theory (which Einstein, Planck, and others would eventually develop) and hardware of his day. Melloni conducted several experiments measuring IR radiation but was vexed by unknown nonlinearities, lack of standard sources, unknown source characteristics, and unknown filter effects. One cannot help but wonder what great discoveries he would have made if he had possessed stable, accurate, and known sources.

It might seem to the modern day laboratory rat that not much has changed. Today, there is more theoretical basis and better, more stable hardware. With substantial effort, the above uncertainties in absolute calibration can be obtained between an instrument and the primary source. In general, the repeatability of the instrument, repeatability of standards, transfer standard errors, and test facility uncertainties induce errors on the order of the above percentages between the primary radiometric standard and the instrument. Although it is theoretically possible to obtain more accurate calibration, Herculean efforts are required to mitigate error at every step in the process.

LABORATORY BLACKBODY ACCURACY

Subject: Radiometrics, test systems, blackbodies

The Rule

When used in a real-world setup, laboratory blackbodies often are accurate to only a few percent.

Basis for the Rule

This rule is based on empirical observations of state of the art. Most commercial blackbodies have a few percent or so variance across their aperture due to (1) reflections, (2) imperfect emissivity, and (3) temperature inaccuracies. Rarely are common commercial blackbodies traceable to a national radiometric standard.

Cautions and Useful Range of the Rule

This is a gross generalization and approximation only; custom-made blackbodies may be better, and many that are actually used are worse.

This is based on commercial blackbodies. Blackbodies employing phase change materials and laboratories that exercise extreme care can claim better accuracy. Conversely, poor radiometric facilities and blackbodies that are damaged or used outside their intended temperature range can be much worse. Radiometric accuracy tends to decrease as blackbody temperature decreases (with some exceptions, such as at the phase change temperature of a cryogen at 4 or 77 K).

Usefulness of the Rule

This rule is useful for underscoring the danger of assuming that a "blackbody" is really "black" and for providing some estimates of the limits of calibration accuracy.

Notes and Explanation

When blackbodies are incorporated into test facilities, several practical constraints limit the radiometric accuracy.

1. There is a small temperature cycle caused by the control bandwidth. This can be on the order of 0.1 percent.
2. A minor temperature uncertainty results from the separation of the emitting surface from the heating and cooling mechanisms and the temperature measurement devices, all contained within the blackbody (but not on the radiating surface). The resultant temperature gradients are small, but small changes in temperature can cause significant changes in radiant exitance. For example, a 1 K bias (at 300 K) alone, causes a 3.7 percent error in the 3- to 5-μm band and a 1.7 percent error in the 8- to 12-μm band.
3. Black coatings are not perfectly black; rarely is their emissivity greater than 0.99. There tends to be a slight emissivity variance across the aperture and a small reflectance.
4. The blackbody may have contaminants on it or be slightly (yet unknown to the user) damaged.

To illustrate possible errors, the plot shown in Fig. 14.2 demonstrates the sensitivity of photon flux to minor changes in temperature, emissivity, and reflection. The plot was made by comparing the calculated photon flux from a perfect blackbody

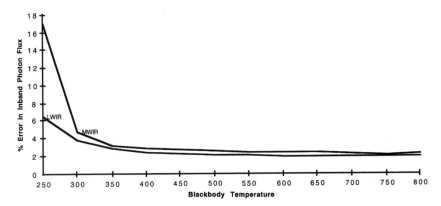

Figure 14.2 Blackbody Accuracy (0.15% temperature error and 0.15% emissivity error)

(emissivity of 1, exact temperature, and no reflection) to an imperfect one (emissivity of 0.9985, 0.15 percent temperature bias, and a reflection of 0.0015 of a 300 K background). MWIR refers to a 3- to 5-µm bandpass, and LWIR is an 8- to 12-µm bandpass.

TOTAL EMITTANCE FROM A BLACKBODY AT 1000 K

Subject: Radiometry, blackbodies

The Rule

A 1000 K blackbody produces 57 kW/m^2 of surface.

Basis For the Rule

This is based on simple radiometry (σT^4).

Cautions and Useful Range of the Rule

The above refers to all wavelengths and applies to blackbodies only.

Usefulness of the Rule

The rule is useful for quick scaling of total emitted power as a function of temperature and area.

Notes and Explanation

Simply applying the Stefan-Boltzmann law ($W = \varepsilon\sigma T^4$), one can calculate the power emitted by a 1 m^2 blackbody as

$$W = (1)(5.6686\times10^{-12} \text{ W cm}^{-2}\text{ K}^{-4})(1000\text{K})^4(100 \text{ cm})(100 \text{ cm})$$

or a whopping 56,686 W. Therefore, a black surface at a temperature of 1000 K produces 56,700 W/m^2. Knowing this, one can scale the watts per meter produced as a function of temperature by T^4. Additionally, emissivity and area scale linearly, so 1 cm^2 of black surface at 1000 K emits 5.7 W.

BRIGHTNESS OF COMMON SOURCES

Subject: Radiometry, phenomenology, environments

The Rules

1. Near the equator, at noon, the sun has a brightness of 10^5 lux.
2. A full moon is 500,000 times dimmer (0.2 lux).
3. A superpressure mercury lamp emits about 250 W/cm² sr.
4. A 60-W bulb emits approximately 50 lux, which is 50 lumens/m² at 1 m.
5. A 4-mW HeNe emits about 1 million W/cm² sr.

Basis for the Rules

The above is based on state-of-the-art technology, basic radiometry, and measurements of natural sources.

Cautions and Useful Range of the Rules

The above rules represent approximations only. They do not include any atmospheric effects, nor do they include spectral effects (e.g., much of the power from a mercury lamps is emitted in a few spectral lines).

Usefulness of the Rules

The rules are useful for quick estimates of the brightness of a source and to develop a feel for the brightness of various sources.

Notes and Explanation

The above rules give the approximate, expected brightness for several common sources of visible light. They also demonstrate the vast differences in brightness between various sources.

It is interesting to compare these brightnesses to the sensitivities of some common visible sensors. Note that a fully dark-adapted eye, under ideal conditions, can produce a visual sensation with about 10^{-9} lux. Since the area of the pupil is 13×10^{-6} m², the eye can see about 10^{-14} lumens, which is merely 30 photons/s. This is confirmed by the Burle *E-O Handbook*,[1] which quotes work showing this sensitivity at 58 to 145 photons per second. Note that only a fraction of these photons actually get to the retina.

Conversely, a standard photomultiplier tube has a dark current approximately equivalent to 5×10^{-10} lumens, or about 1.5 million photons/s (again, as with any rule of thumb, some may be different; e.g., S-24 at 1×10^5 e⁻/s).

In addition, most photographic film responds if 10^{-2} lux is imposed for 1 s. At a film resolution of 30 lines/mm, we could see the effect on 10^{-9} m² or with 10^{-11} lumens (about 30,000 photons/s for 1 s = 30,000 photons).

Reference

1. Burle Industries. 1974. *Burle Electro-Optics Handbook*. Lancaster, PA: Burle Industries, 46.

Other Source

M. Klein. 1970. *Optics*. New York: John Wiley & Sons, 129.

THE RULE OF 4f#^{2}

Subject: Radiometry, systems, optics

The Rule

From an extended background, the scene power on a detector can be

$$E_d \propto \frac{M}{4\,(f\#)^2}$$

where E_d = irradiance (power) on the detector
M = radiant exitance (power) from the scene
f# = effective f# (effective focal length/effective aperture)

Basis for the Rule

This rule is based on basic radiometry and is an outgrowth of associated rules in this book. It can be shown that, for an extended object (e.g., the background), the power on the detector depends only on the f# and not the aperture size.

Consider that the energy (in watts) at the aperture is as follows:

$$\text{Energy at aperture} = \frac{L\pi D^2}{4R^2}\,(IFOV)^2\,(R)^2$$

$$= \frac{LD\pi}{4}\left(\frac{\text{detector dimension}}{f}\right)^2$$

$$= \frac{M}{4\,(f\#)^2}\,(\text{detector dimension})^2$$

where R = range to the target
D = diameter of the receiving aperture
L = radiance of the background
f = focal length of the optics

Cautions and Useful Range of the Rule

The above relationship ignores atmosphere and optical transmission effects. It is valid for extended sources (not at all valid for point sources).

Usefulness of the Rule

This handy rule allows easy use and estimation of the relationship between the power from the scene and the expected power on the detector. It also provides for calculation of the effect of various f#s on detector power (e.g., will the new f# cause the detector pixel's wells to overfill?).

Notes and Explanation

An extended source will result in a flux on the detector that is governed by the f# of the optical system only. Again, the actual calculation will need optical transmission, blur spot effects, and consideration for spectral filtering.

SOLAR REFLECTION ALWAYS ADDS TO SIGNATURE, OR RA'S RULE

Subject: Radiometry, contribution of solar irradiance to target signature

The Rule

The sun is bright. When present, solar reflection always adds signature to a target, regardless of the bandpass used to view it and the background. Specific gains depend on the conditions, sensor, and bandpass.

Basis for the Rule

This rule is based on empirical arguments about solar scattering and calculations of blackbody emissions.

Cautions and Useful Range of the Rule

Increases in signature from solar reflection are usually great for wavelengths less than ≈3 μ and inconsequential beyond ≈5 μ. Solar reflection is usually measurable but not significant in between the two.

Atmospheric effects must be considered if operating within the atmosphere.

Usefulness of the Rule

This rule is useful to consider in system-level trade-offs and for the analysis of designs driven by target signatures.

Notes and Explanation

In the visible and UV spectral regions, objects are typically viewed by solar reflection only. In the infrared, the object's own thermal emission is normally used to provide the signal to detect the target. However, in the short-wave and mid-wave infrared, reflection of radiation emitted by the sun (and sometimes the earth or moon) can contribute significantly to a cold object's signature. The contribution may be enough to allow smaller optics or a less sensitive focal plane.

Example

We want to observe a 1 m² satellite at a temperature of 300 K, a reflectivity of 0.3 (Lambertian), and an emissivity of 0.7 with an 8 to 12 micron, a 3 to 5 micron, and a visible bandpass. The observation is done by another satellite in earth orbit. By using the Planck equation and solar emission tables as found in any book on IR sources, the following table can be generated.

	0.4–0.6-μm band	3–5-μm band	8–12-μm band
Thermal radiant intensity (W/sr)	essentially 0	1.5	28
Solar reflection (W/sr)	35	2.2	0.7
Total (W/sr)	35	3.7	29
Percent of signature contribution by the sun (%)	100	60	2

We can see that the solar contribution is dominant in the visible. In the IR bands it contributes significantly to the signature of the MWIR band while only a minor contributor to the LWIR band. Nevertheless, it does contribute something to every band.

Bandpass Optimization

Subject: Radiometry, phenomenology, sensor design, bandpass optimization, thermal emission

The Rule

For maximum performance in the thermal infrared, the bandpass should include the maximum change in photon emission with temperature or

- for a BLIP case, the bandpass should include the maximum of the derivative

$$\frac{\partial Q_\lambda}{\partial T \sqrt{Q_\lambda}}$$

where Q_λ = photon exitance
T = temperature

- for a non-BLIP case, the bandpass should include the maximum of the derivative

$$\frac{\partial Q_\lambda}{\partial T}$$

Basis for the Rule

This rule is based on simplification of radiometrics as described below and empirical manipulation of the Planck function.

Cautions and Useful Range of the Rule

This rule assumes that the performance is driven by the temperature differential. Therefore, this rule does not apply to detecting targets via reflected sunlight (usually in bandpass below 3 µm), spectral emitters, or chemical emissions (engine plumes).

This does not account for technology and application limitations. This rule may indicate that you should have a bandpass that includes 13 µm, but finding a large - scale focal plane that responds to that wavelength is difficult.

The rule does not account for atmospheric effects (sometimes mean, old Mr. Atmosphere does not allow you to use the best band) or differential emissivity effects.

A composite bandpass is usually best for multiple targets. That is, a target class that may have targets of different temperatures should include the maximum derivative for all targets.

Multiple bandpasses are always better (allowing multispectral or hyperspectral algorithms) and are not addressed by this rule.

Unfortunately, if the target and background are close to the same temperature, this rule will likely maximize clutter as well. A separate signal-to-clutter analysis should be performed.

The above equations and the plot are for detectors that sense "photons," not watts. This includes photoconductive and photovoltaic semiconductors, Schottky barriers, and quantum wells. However, this does not apply to bolometers, which respond to the power placed on them. For this type of detector, the rule should be

- The bandpass should include the maximum of the derivative

$$\frac{dW_\lambda}{dT}$$

where W_λ = differential radiant exitance

Usefulness of the Rule

This rule allows for a first estimate at determining the optimal bandpass for a blackbody or graybody target. It allows a quick determination of bandpass and limits bandpass trade-offs.

Notes and Explanation

Bandpass selection to maximize performance is often a trade-off between maximizing the absolute contrast and maximizing the signal-to-noise (or signal-to-clutter) ratio. Usually, with a single-bandpass system, maximum performance is achieved when the bandpass includes the fastest changing part of the photon exitance (emission) as a function of change in temperature. If the background and target are close to the same temperature, reflectivity, and emissivity, then you want to set the bandpass to maximize the derivative of the maximum photon flux per wavelength. If the target and background are different in temperature, then the bandpass should be set to include the maximum photon emission from the target. Interestingly, for thermal emitters, this maximum usually starts just shy of the Planck function peak.

Most infrared sensor systems are built to detect a target via the difference in photon emittance caused by a slight change in temperature between it and its background. According to Planck's theory, for a given temperature, there is a unique spectral point where a small change in temperature results in the largest change in radiant exitance. If possible, the bandpass should include this wavelength. For instance, from Fig. 14.3, one can see that, if the target is at 320 K, then the maximum change in radiant exitance for a system that is *background limited in performance* (BLIP) occurs near 7.6 microns. The sensor should be designed for a bandpass that includes 7.6 microns. However, this is a poor wavelength for atmospheric transmission, so a sensor operating in the atmosphere should try to get as close as possible and still be in a transmission region. The dark side is that the clutter effects also tend to be the greatest at the wavelength where these derivatives are maximum for the temperature of the background.

The figure shows two curves plotted by temperature. The top curve is the maximum of the partial derivative of

$$\frac{\partial Q_\lambda}{\partial T}$$

that corresponds to a case where the noise is dominated by conditions independent of the background or target temperature. For this case, the bandpass should, if atmospherics, technology and design constraints permit, include this maximum. The lower line is for a BLIP case, and it plots the maximum of

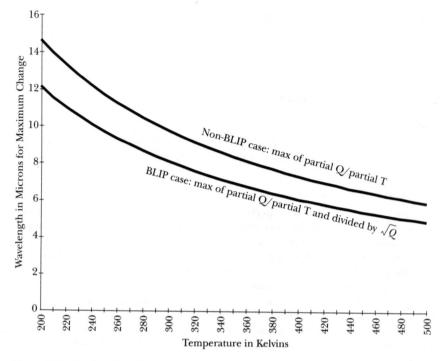

Figure 14.3 Wavelength vs. Temperature (courtesy of Dr. George Spencer)

$$\frac{\partial Q_\lambda}{\partial T \sqrt{Q_\lambda}}$$

This is the same as the non-BLIP case, except it is also divided by the square root of the photon exitance from the target (or background). A BLIP case would have the photon arrival variations dominating the noise, so it can be included easily.

Reference

Rule partially developed by Dr. George Spencer, 1995.

THE ETENDUE, OR OPTICAL INVARIANT, RULE

Subject: Optics, radiometry, systems

The Rule

At any plane in a lossless optical system, for a given wavelength, the product of the solid angle and the area of the ray bundles that the radiation is traveling through is a constant, or

$$C = A_o \Omega$$

where C = a numerical constant for a given detector pixel size and wavelength
  A_o = useful area of the optics
  Ω = solid angle field of view

Additionally (with reference to Fig. 14.4 and the explanation in the "Basis of the Rule" section), we can write:

$$A_d \Omega_i = A_s \Omega_o = A_o \Omega' \approx A_o (\text{IFOV}) \approx C_d \lambda^2$$

where A_d = area of the entire detector (or pixel in an array)
  Ω_i = solid angle subtended by the detector in image space
  A_s = area of interest in the scene
  Ω_o = solid angle of the scene in object space
  A_o = area of the optics
  Ω' = solid angle of the detector projected into object space
  IFOV = instantaneous filed of view of a detector (FPA) pixel
  λ = wavelength (for a broad bandpass system, use the midpoint for this)
  C_d = a constant determined by the pixel geometry's relationship to the blur diameter (see below). Generally, for imaging systems, this is from 1.5 to 10, although it may be higher for systems in which it is advantageous to oversampling the blur, such as star trackers

Basis for the Rule

The Etendue relationship is a basic premise of radiometry and optics. It derives its foundation from the conservation of energy, diffraction theory, and thermodynamic principles applied to optics.

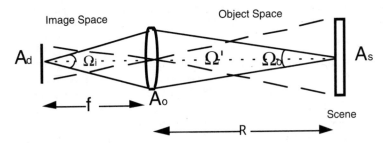

Figure 14.4 Geometric Definitions

In a diffraction-limited system, the blur diameter is equal to the Airy disk or 2.44 $(\lambda/D)f$, where D is the aperture diameter and f is the focal length. If a square detector is matched to the blur, its area is the square of this or $5.95(\lambda^2/D^2)f^2$. The solid angle seen by the detector is A_o/f^2, so the product of

$$A_d\,\Omega_i \approx 5.95\left(\frac{\lambda^2}{D^2}\right)(f^2)\left(\frac{A_o}{f^2}\right) \approx 6\lambda^2$$

In systems that are not diffraction limited, the "6" (or C_d) is replaced by a larger number, but the important λ^2 dependence remains. Conversely, if the blur spot is oversampled (large compared to the detector pixel size), this will be smaller (e.g., $1.5\,\lambda^2$ if the radius of the Airy disk is matched to that of the detector pixel).

Similarly, a pixel (again, matched to the Airy disk) projected onto the scene has an area of

$$\left[2.44\left(\frac{\lambda}{D}\right)R\right]^2$$

where R is the range, and the solid angle is

$$\frac{\left[2.44\left(\frac{\lambda}{D}\right)R\right]^2}{R^2} \text{ or simply } \frac{5.95\lambda^2}{D^2}$$

and, for an unobscured circular aperture,

$$A_o\,\Omega' \approx \left(\frac{5.95\lambda^2}{D^2}\right)\left(\frac{\pi D^2}{4}\right) \approx 4.7\lambda^2$$

Cautions and Useful Range of the Rule

The rule assumes a lossless system; actual engineering calculations should include some transmission loss.

The rule does not account for changes in index of refraction.

The generic Etendue works regardless of aperture and pixel shape (it is frequently used by spectroscopists working with slits); note that the basis for parts of the second equation assume an unobscured circular aperture and a square pixel matched to a circular blur.

Usefulness of the Rule

The rule is a fundamental premise of much of geometrical optics and radiometry. When properly applied, it allows one to estimate another (e.g., a competitor's) system's useful aperture, f#, or detector size. (If you do not know the speed of the system, assume an f# 1; rarely are faster speeds achieved.) It provides a determination of the collection aperture for radiometric applications.

It provides a quick calculation of any of (a) the wavelength, (b) aperture size, or (c) IFOV, if the other two are known.

It can be used for estimates when coupling light into fibers, because there is a small cone, defined by the equation, that allows acceptance of light into a fiber.

Notes and Explanation

This rule goes by many different names. Spectroscopists like *Etendue,* ray tracers like *Lagrange theorem,* and radiometry buffs like to say *optical invariant.*

The importance is that the useful aperture and the field of view are inversely related to each other. The numerical value of their actual relationship depends on the optical design, the chosen paraxial ray, and the height of the object or detector. The numerical constant is not important—you can choose that based on your design or assumptions. What is important is the understanding that increasing one must result in the decrease of the other. Hence, large (multiple-meter) astronomical telescopes have small fields of view, and wide-angle warning systems have very small apertures. You can have one, but not the other—yup, life is tough. Additionally, a zoom lens will have a larger effective aperture when viewing its narrow field (resulting in a brighter image) as compared to its wide field.

As Longhurst[1] puts it, "In paraxial geometrical optical terms, the ability of an optical system to transmit energy is determined by a combination of the sizes of the field stop and the pupil in the same optical space; it is measured by the product of the area and the pupil in the same optical space; it is measured by the product of the area of one and the solid angle subtended at its center by the other. This is the three-dimensional equivalent of the Helmholtz-Lagrange invariant or the Sine Relation."

Given the size of a detector and the loose approximation that it can "sense" energy from one steradian of solid angle, the upper limits of the field of view and capture area in any associated optical system are immediately calculable. The energy captured (using a small angle approximation) is approximately Ω_o or A_o/R. For a "fast" system, this is on the order of unity, so the energy captured is approximately A_d/A_o. In any event, a small A_d implies small capture area for a given FOV (hence, low energy capture). A large IFOV implies a small capture A_o area for a given detector area (A_d).

Reference

1. R. Longhurst. 1976. *Geometrical and Physical Optics.* New York: Longman, 465–467.

Additional Sources

Rule provided by Dr. J. Richard Kerr, 1995, and partially developed by Dr. George Spencer, 1995.

R. Kingslake. 1988. *Optical Systems Design.* Orlando, FL: Harcourt Brace Jovanovich, 36–38, 43–44.

A. Siegman. 1986. *Lasers.* Mill Valley, CA: University Science Books, 672.

I. Taubkin et al. 1994. Minimum temperature difference detected by the thermal radiation of objects. *Infrared Physics and Technology* 35 (5), 718.

C. Wyatt. 1987. *Radiometric System Design.* New York: Macmillan, 36, 52.

15

Shop Optics

The world is full of rules of thumb for the shop manufacture and test of optics. Included here are a select few that should be available to anyone in electro-optics.

Shop optics began many centuries ago, when the first optical hardware was made. To early investigators, the world was full of rules of thumb and principles explained by thought process alone. Certainly, shop optics developed earlier than 425 B.C., when the first known written account of making lenses was written (*The Clouds,* by Aristophanes). Euclid, in his 300 B.C. publication *Optics,* may have been the first to write about curved mirrors, and by the eleventh century, glassblowers were making spectacles and lenses for research.

Clearly, optical shop manufacture and test was flourishing shortly after 1600, with the invention of both the telescope and microscope and the popularity of spectacles. The publication of Isaac Newton's *Opticks* in 1704 was a milestone in the science and engineering of optics. His work was a compendium of theories to explain the optical phenomena that he observed in his optics shop.

Jean Foucault (1819–1868) gave mirror makers another powerful tool called the *knife edge test.* A straight edge, often a razor blade, is used to block the light at the focus of an optic. The appearance of the mirror, when viewed from near the knife edge, is a powerful indication of the figure of the optic. However, you should be careful not to nick your nose while performing this test.

Key to modern shop optics are modern materials and testing techniques. Many very precise testing techniques use interferometers. Hence, many of the following rules address fringes—the light and dark patterns produced by interferometers. The interferometer is one of the most accurate and finest measurement devices known to the human race, with measurement accuracy of better than half a millionth of a centimeter (with a tenth of a wave readily achievable). Interferometers can easily display wavefront deformation much smaller than a wavelength, so they are often used to measure the shape of an optical surface. The Fizeau interferometer is one of the commonest for optical metrology. Usually, the optical surface under test is compared with a "known" flat surface. The light waves reflected from the two surfaces produce interference fringes whose spacing indicates the separation between the two surfaces. If the reference surface is truly flat, the fringe spacing directly indicates the figure (or shape) of the test surface.

SHARP AND SYMMETRIC FRINGE CRITERIA, OR TOLANSKY'S INEQUALITY

Subject: Shop optics, multiple-beam interferometers

The Rule

To form high-quality, sharp, and symmetric fringes, the following restriction should apply:

$$\frac{8}{3} n^3 \varepsilon^2 t < \lambda$$

where n = number of interfering beams
    ε = "wedge" angle between the two beams
    t = separation between the plates
    λ = wavelength

Basis for the Rule

This relationship is derived from Fizeau fringe theory and empirical observations.

Cautions and Useful Range of the Rule

As with any interferometric measurement, the above rule is sensitive to the specifics and subtitles of the test setup.

The above is valid for monochromatic light and does not hold true for white light.

Usefulness of the Rule

The above is useful to check and ensure the visibility of fringes. As such, it is also useful when testing optics.

Notes and Explanation

If the above inequality becomes an equality, then the nth beam will add destructively rather than constructively. If the surfaces are in contact, the plate separation is usually driven by dust particles on the order of 1 μm, so t will be about 1 μm. The limit of the angle ε is driven by the number of fringes one wishes to view (the smaller the angle, the larger the spatial separation between two adjacent fringes). High-quality, high-finesse fringes require a superposition of a large number of regularly delayed beams. The number depends upon the reflectance of the surfaces but should be near 100 for a reflectance of 0.92, 200 for 0.96, and 300 for 0.98. For reflectances below approximately 0.9, the number of interfering beams needed is roughly twice the required finesse.

Malacara[1] gives the original reference for this as S. Tolansky. 1948. *Multiple Beam Interferometry of Surfaces and Films.* Oxford, U.K.: Oxford University Press.

Reference

1. D. Malacara. 1978. *Optical Shop Testing.* New York: John Wiley & Sons, 186–189.

SCRATCH AND DIG

Subject: Shop optics, optics, materials

The Rule

Commercial-grade IR optics usually have a scratch-and-dig specification of about 80/50 to 80/60. The specification is usually 60/40 for better-quality visual optics, 40/20 for low-power active applications, and 10/5 for high-power lasers.

Basis for the Rule

This rule is based on empirical observations and simplification of scatter theory as outlined in the U.S. military specifications. This is defined in three U.S. military specifications: Mil-O-13830A, C7641866, and Mil-M-13508C.

Cautions and Useful Range of the Rule

Generally, this is valid for the generic uses outlined above but, remember, it is a rule of thumb.

Some high-resolution, low-scatter optics will require more stringent specifications. Conversely, some low-cost, high-volume production applications may require less stringent specifications.

Surfaces near images (e.g., reticles) require a finer scratch and dig specification. Digs of less than 2.5 μm usually are ignored for IR applications.

Usefulness of the Rule

The rule is helpful in understanding the scratch-and-dig specifications that are appropriate for various applications. It is also useful for generating the specification and to get a feel for what to expect from a vendor of a given quality.

Notes and Explanation

When manufacturing an optical surface, various grit sizes are used. Unfortunately, it is unavoidable to have grit of one size contaminate the polishing from grit of a smaller size. This leads to "scratches" and "digs" on the optical surface. When these scratches are small enough and few enough, no noticeable degradation in spot size or MTF is noticeable. It is therefore wise (to keep costs down) to specify the scratch and dig at a level where the errors are slightly less than expected from diffraction and aberrations. The rule provides the current conventional wisdom on the level needed.

The first number is the "scratch," and the second is the "dig." These refer to two graded sets of surface quality standards drawn in military standard C7641866. The units to the scratch and dig are normally excluded, for some reason. They happen to be in units of 1/100 mm for dig. The scratch is a unitless ratio. It is defined in MIL-O-13830A as follows: "When a maximum size scratch is present, the sum of the products of the scratch numbers times the ratio of their length to the diameter of the element or appropriate zone shall not exceed one-half the maximum scratch number."

Sources

Rule partially developed by Tom Roberts, 1995.
J. Miller. 1994. *Principles of Infrared Technology.* New York: Van Nostrand Reinhold, 64.
United States Military Specification MIL-O-13830A.

PRODUCIBLE LENS RULES

Subject: Shop optics, optics, cost, manufacturability, producibility

The Rule

To facilitate lens manufacture and mounting, the former Soviet optical industry recommended that the lens diameter (D), axial thickness (d), and edge thickness (t) be related as follows:

1. for positive lenses, $4d + 10t \geq D$, with $t \geq 0.05D$
2. for negative lenses, $12d + 3t \geq D$, with $d \geq 0.05D$

Basis for the Rule

This rule is based on empirical manufacturing data in the former Soviet Union.

Cautions and Useful Range of the Rule

Clearly, this rule depends on the maturity of the technology. Moreover, it does not apply to small wide angle or small microscope lenses. Also, it does not address material costs or complexities, so it assumes readily available materials.

Usefulness of the Rule

It is useful as a quick check to verify that the lens that you are specifying is manufacturable and producible.

The former Soviet optics industry is aggressively marketing their technology to the West. They have used such a rule in the past to gauge producibility, so it is wise to employ it when dealing with them now. Moreover, this is a valid rule for simply ensuring easy producibility of low-cost optics. If low cost or volume producibility is required, then it is useful for any optical engineer to compare each of the elements to ensure this is met (regardless of the source of the actual "glass").

Notes and Explanation

Because of the mechanical properties of typical lens materials and the manufacturing process, it becomes difficult to manufacture lenses when the prescription requires them to be very thin. The above rule ensures that the lenses will be manufacturable in quantity, at low cost.

Sources

Various Soviet lens design standards.

B. Begunov, N. Zakaznov, S. Kiryushin, and V. Kuzichev. 1988. *Optical Instrumentation, Theory and Design.* Moscow: MIR Publishers, 73–74.

MOUNTING AND CENTERING DIFFICULTIES

Subject: Shop optics, optics, cost

The Rule

The process of mounting and centering the optical element is the most difficult part of the manufacturing process.

Basis for the Rule

This is based on empirical observations, state-of-the art technology, and common grinding and polishing procedures.

Cautions and Useful Range of the Rule

Manufacturing technology changes. With time, this rule may be invalidated.

Usefulness of the Rule

The rule underlies a basic producibility and cost driver. An optical element's cost is greatly affected by the tolerance on the centering and the difficulty of mounting.

Notes and Explanation

The tedious process of mounting a lens or mirror and centering the element is one of the most critical and time-consuming parts of the manufacturing process. This is because of the enormous precision typically demanded by the specifications and the way optics are made. Whether we employ a old-fashioned lapping process or more modern diamond turning, the initial placement of the optic to be ground determines the final shape and accuracy of the grinding process. The actual grinding is mechanical and can be done when everyone is at home sleeping, but the placement and final centering require much touch-labor.

Source

R. Kingslake. 1988. *Optical Systems Design.* Orlando, FL: Harcourt Brace Jovanovich, 4.

FRINGE MOVEMENT

Subject: Shop optics, testing optics, optics

The Rules

When testing an optic in an interferometer:

1. If you gently press (e.g., with pencil eraser) on the edge of the upper optic and the fringes move toward the point of pressure, then the upper surface is convex; conversely, it is concave if the fringes more away from this point (due to the differences between the two surfaces).

2. If you press near the center of the fringe system on the top optic and the surface is convex, the center of the fringe will not change, but the diameter will increase.

3. If the source of the light is white light, and pressure is applied to the convex center, the first fringe will be dark, and the first light fringe will be white. The next fringe will be tinged bluish on the inside and reddish on the outside. A concave surface will have the outer fringe as the dark one, and the color tingeing will be reversed.

4. When fringes are viewed obliquely from a convex optic, the fringes appear to move away from the center as the eye is moved from normal to oblique. The reverse occurs for a concave surface.

Basis for the Rules

The rules are based on an analysis of the optical path difference for a Newton interferometer to achieve good contrast.

Cautions and Useful Range of the Rules

These relationships assume that the test optic is referenced to a standard flat, and that the measurements are made in air with the air gap less than 6λ (the surface under test fits well into the test plate).

Usefulness of the Rules

This is useful for testing optics and understanding the fringe patterns.

Notes and Explanation

This is based on analysis of the optical path difference caused by a varying air thickness between the optic and test flat. For instance, if you apply pressure near the center of a concave, the air is forced out, leaving a smaller optical path difference.

Source

D. Malacara. 1978. *Optical Shop Testing.* New York: John Wiley & Sons, 8–11.

Approximations for the Foucault Knife Edge Tests

Subject: Shop optics, testing optics, optics

The Rules

A Foucault knife edge test indicates

1. spherical aberration, if the shadow shows more than one dark region

2. coma, if the shadow pattern consists of rectangularized hyperbolas or an ellipse

3. astigmatism, if the shadow is a straight line with a slope and can be made to rotate by using different placements of the knife edge about the optic axis

Additionally, the center of curvature or a defocus error can be tested by observing the pattern as the knife is moved in relationship to its distance from mirror. If the knife is placed first inside the focus then outside the focus, the shadow will change sides.

Basis for the Rule

This is based on diffraction theory and empirical observations of the patterns from a knife edge test.

Cautions and Useful Range of the Rule

The knife edge test is sensitive to knife edge placement and test setup.
    Astigmatism may escape detection if only one knife orientation is used.

Usefulness of the Rule

This is useful for quick testing of optics or observing Foucaultgrams.

Notes and Explanation

Foucault's knife edge test consists of cutting off the image of a point source with a straight edge and observing the apparent shadows on the mirror. This test is a quintessential optical shop test for transverse aberrations. It is useful for measuring the radius of curvature and as a null test. It is easy to implement, and the experienced optician can derive a wealth of knowledge about the surface being tested. The accuracy of the Foucault knife edge test can be impressive. It has been estimated that with the eye alone (assuming a 2 percent contrast) that wavefronts can be tested to $\lambda/600$. It is frequently employed as an in-process test to determine the status of the figure for additional polishing.

When conducting the knife edge test and viewing the shadows (often called a *Foucaultgram*), they look like a very oblique illumination of the surface. If the source is to the right of the knife edge, the apparent illumination is from the right. A lamp on the surface looks bright toward the light (right in this case) and dark on the other side. A divot looks dark on the right and light on the left. If the wavefront is perfectly spherical, the Foucaultgram looks a uniform gray. Some commonly observed imperfection phenomena include:

1. The lemon peel, which indicates local roughness and poor polishing

2. A bright edge toward light, which indicates a turned-down edge

3. A dark edge toward light, which indicates a minor miracle—a turned-up edge

Sources

D. Malacara. 1978. *Optical Shop Testing*. New York: John Wiley & Sons, 231–253.
Other information supplied by W. M. Bloomquist, 1995.

DETECTION OF FLATNESS BY THE EYE

Subject: Shop optics, optical testing, ergonomics

The Rule

The naked eye can detect a lack of flatness having a radius of curvature up to about ten thousand times the length of the surface being viewed.

Basis for the Rule

This is based on empirical observations of optical testing and ergonomics.

Cautions and Useful Range of the Rule

The surface must be at least several times the diameter of the eye's pupil (e.g., 50 mm or more), and the above assertion assumes a normal, healthy human eye.

Usefulness of the Rule

This rule comes in handy when inspecting an optic.

Notes and Explanation

The test of a flat surface by oblique reflection is so sensitive that even the naked eye will detect quite a low degree of sphericity, if nearly grazing incidence is employed. For instance, if a dotted cross is deployed and an image is viewed, then any astigmatism resulting from the spherical form of the surface will cause the dots of either the vertical or horizontal lines to merge together into a gray line.

Source

B. K. Johnson. 1960. *Optics and Optical Instruments*. New York: Dover Publications, 196–197.

COLLIMATOR MARGIN

Subject: Shop optics, optics

The Rule

It is good design practice to make the diameter of the collimating mirror in a test setup at least 10 to 20 percent greater than that of the optics to be tested, and to make the focal length at least 5 to 10 times that of the element under test.

Basis for the Rule

This rule is based on experience in the school of hard knocks of optical testing. It eases the difficulty of aligning and placing optics in the test setup.

Cautions and Useful Range of the Rule

Some optics that are sensitive to extreme off-axis radiation will need larger collimators.

Under some highly controlled conditions, accurate measurements can be taken with a collimator only slightly larger than the entrance optics.

In a really tough situation, the collimator can be undersized. If so, much data must be taken, and the information (interferograms) can be spliced together. However, this requires that one pay consideration to schedule and budgets.

Since off-axis collimators must turn the beam, they need to be even larger.

Usefulness of the Rule

Adhere to this rule whenever using a collimator; otherwise, you will spend many hours hunched over an optical table wondering why it is not working right. It should be considered whenever you perform test system design, determine collimator specifications, collect test fixtures, and so on.

Notes and Explanation

Generally, it is wise to have the test apparatus larger and more accurate that the item to be tested. The collimator's useful exit diameter should be significantly larger than the diameter of the element under test so that placement of the lens is not critical (10 to 20 percent minimum). This also allows for some slop in pointing, ensures that the optics under test will be completely filled, and reduces the deleterious contributions from off-axis sources. Moreover, it is good design practice to make the focal length of the collimating mirror 10 times the focal length of the lens under test.

Source

Rule partially developed by Max Amon, 1995.

Cleaning Optics Caution

Subject: Shop optics, optics, systems

The Rule

Dirty optics should be cleaned only after great deliberation and with great caution.

Basis for the Rule

This rule is based on empirical observations. Wiping dirt off often makes the surfaces more defective that the dirt did. This is because

1. Most surfaces, and all fingers, have very fine abrasive dirt on them that will scratch an optical surface.
2. A few big areas of dirtiness are less harmful (scatter less light) than the myriad long scratches left behind after removing the hunks.
3. Small particles can adhere very strongly (in proportion to their mass) and cannot be blown or washed off easily.
4. Washing mounted optics just moves the dirt into mounting crevices where it will stay out of reach, waiting for a chance to migrate back to where it is harmful.

Cautions and Useful Range of the Rule

Sometimes it is necessary to clean optics, especially when the contaminant is causing excessive scatter or if a laser will burn them into the coating.

The longer the wavelength, the more valid this becomes (e.g., UV systems usually need to be cleaner than IR systems). Additionally, optics near a focal plane need to be cleaner than optics located at the aperture.

Usefulness of the Rule

This is something to remember before you reach for that paper towel, and it underscores and explains a basic good idea.

Notes and Explanation

It is often surprising how well optical systems work when they are dirty. There are legends about tactical FLIR systems working fine on a mission. When inspected afterward, the crew is surprised to find the window splattered with mud and dead bugs. Additionally, many older telescopes (e.g., the Palomar's 200-inch primary) have numerous surface nicks, cracks, and gores, and yet seem to work fine (once they are blackened out). Most observatories wait until a few years worth of dust accumulates on their primaries before washing them. Again, optics near a focal plane (e.g., reticles, field stops, field lenses) must be kept cleaner. This is because a particle or defect on a surface near the focal plane, projected back to the front aperture, could be a large portion of the collecting aperture.

There are several reasons for this apparent contradiction. First is that the human eye is especially good at seeing imperfections (dirt, pits, and so forth) on smooth surfaces. Second is that the dirt usually does not amount to a large fraction of the surface area, so the diffraction, MTF, and transmission losses are surprisingly small. Third is that these particles are far out of focus. Often, the most important effect of the dirt is scatter.

Optics should be stored in containers under a laminar-flow bench. When is use, hanging upside-down helps. When you do clean, be very careful of the water and cloth that you use. Soaps often have some sandy grit in them to add friction for easier dirt removal. Additionally, alcohol-based perfumes are frequently added to cleaning products and may remove optical coatings.

Source

Rule provided by W. M. Bloomquist, 1995.

ACCURACY OF FIGURES

Subject: Shop optics, optics

The Rule

The *Rayleigh quarter-wave rule* states that an optic whose output wavefront is accurate to a quarter of a wavelength is good enough for most applications.

Additionally, a figure of $\lambda/15$ is required for interferometric, phase sensitive, and critical imaging applications; while a $\lambda/10$ is often sufficient for imaging optics requiring low beam distortion (especially with multiple elements), and merely $\lambda/4$ is required for general purposes.

Basis for the Rule

This rule is basically based on the simplification diffraction theory and Strehl ratios as applied to electromagnetic theory and backed by empirical observations.

Cautions and Useful Range of the Rule

This refers to the maximum or peak to valley wavefront error.

This is valid for normal applications. Super-high-power lasers, extremely wide fields of view, and other exotic implementations may require more stringent wavefront control.

The optic should be measured at the wavelength of interest (thus, this is a more difficult specification to meet for a UV system than for an IR system).

Usefulness of the Rule

The rule provides a quick estimate of the figure required for a given optical element, based on application.

It is a useful criterion for estimating the amount of allowable aberrations in a typical image-forming system.

Notes and Explanation

An optical surface's *figure* is the shape of the surface. Its quality is measured as a fraction of the wavelength (of intended use) departure from an ideal desired surface. It is usually quoted as a peak-to-valley measurement. Sometimes it is quoted as a root mean square (RMS) wavefront error, which is smaller than a peak-to-valley measurement by a factor of approximately 4. In any event, when someone specifies "λ/X," the larger the denominator, the smaller the departure from the ideal and the better the quality. The appropriate quality requirement depends on its final use. One can afford to be sloppier with the figure for plastic toy binoculars than for an interferometer or a space telescope.

Rayleigh found that, when an optical element has spherical aberration to such an extent that the wavefront of the exit pupil departs from the best fit by $1/4$ of a wavelength, the intensity at the focus is diminished by 20 percent or less. He also found that this could be tolerated and was difficult to notice. Subsequent workers also found that when other common aberrations reduce the Gaussian focus intensity by about 20 percent or less, these too had little overall effect on image quality.

When manufacturers invoke this rule to describe their product, they usually mean $1/4$ of the HeNe laser line at 0.6328 μm; however, the wise engineer will find out exactly at what wavelength the measurements were taken and if it is quoted as peak to valley, peak to peak, RMS, or whatever.

Incidentally, for transmissive optical elements, surface figure error can also be represented as:

$$\frac{WFE}{(\Delta n - 1)}$$

where Δn = index change across the surface
WFE = wavefront error

Source

Information supplied by W. M. Bloomquist, 1995, and Tom Roberts, 1994.
M. Born and E. Wolf. 1980. *Principles of Optics*. New York: Pergamon Press, 408–409.
J. Miller. *Principles of Infrared Technology*. New York: Van Nostrand Reinhold, 64.

ACCURACY OF A FIZEAU INTERFEROMETER

Subject: Shop optics, testing optics

The Rule

The Fizeau interferometer's accuracy is generally one fringe or half of the wavelength. It is possible, with great care, to achieve an accuracy of $\lambda/10$ visually. By using photographs, $\lambda/20$ is usually the best accuracy achievable from shop interferometers

Basis for the Rule

This is founded on empirical observations and general limits of common equipment.

Cautions and Useful Range of the Rule

This rule assumes normal shop interferometers and full circular fringes. Sometimes this accuracy may be exceeded, with great effort.

Usefulness of the Rule

These approximations are useful to keep in mind when attempting to test optics, write a specification, or design a test facility.

Notes and Explanation

When observing fringes, it is possible to determine accuracy to far less than a wavelength. However, there are limits to estimating the deviations in distance from one fringe to another. This rule provides a good guesstimate of the available accuracy. To obtain higher accuracy when photographing the fringes, the camera may be set to a distance of at least >5 diameters of the reference flat.

Accuracy beyond the above limit requires careful calibration of the reference flat, air flow, and camera.

Source

D. Malacara. 1978. *Optical Shop Testing*. New York: John Wiley & Sons, 8–12.

OVERSIZING AN OPTICAL ELEMENT FOR PRODUCIBILITY

Subject: Shop optics, optics

The Rule

A lens' or mirror's radius should be oversized by 1 to 2 mm to ease the grinding of good figure and coating.

Basis for the Rule

This is based on empirical observations of cost and producibility.

Cautions and Useful Range of the Rule

Being based on state-of-the-art and common grinding and polishing procedures, it may be invalidated with time and the march of technology.

Of course, smaller margins can be accommodated for extra cost and risk.

This assumes that the optical piece can be mechanically supported by the extra 1 or 2 mm. Large, heavy optics will require more.

Usefulness of the Rule

This is useful when determining a useful specification for optics that actually can be made. It should also be considered when doing mechanical layouts (e.g., that cold filter in the dewar is actually a little larger than the useful diameter).

Notes and Explanation

Like it or not, an optical element must be handled during manufacture. Often, the least expensive way to handle and mount the element is by the edges. This implies some (small but finite) region to clamp, support, and secure the element. Additionally, every optical element eventually must be mounted into a telescope. Allowing a millimeter or two of slop for glue or encroachment of mounting fixtures eases the building of the entire optical assembly as well.

There are still more reasons to slightly oversize the optics. A small chamfer is needed, or edge chipping is almost inevitable. This requires a little extra space. Additionally, rays close to the edge are also close to becoming "stray light" by scattering from the cylindrical edge of the lens, so it is wise to mechanically block off the edges anyhow. Most modern diamond turning machines and coating apparatus can ensure proper specification to within a millimeter or two of the edges. However, within a millimeter or two of the edges, the figure is usually poor.

16

Systems

The ultimate goal of most EO research and development is to create systems that generate quality information. Systems represent the integration of myriad optical, structural, electronic, and mechanical systems into a group of things that work together to perform a function. As we see in this set of rules, there are some key and underlying factors that can be relied on to aid the system design and analysis process.

EO systems experienced little development prior to WWII. This is probably owing to the immature component technologies and the fact that scientific instruments seem to concentrate on film technology. Film technology was not well suited for television, and visible electro-optical systems were always pushed by industry. During WWII, Britain and Germany attempted to develop useful EO systems. Although significant groundwork was laid, noting much came of these efforts as compared to work in radar. As such, electro-optics did not play a part in that war. In fact, it was not until the Vietnam War that electro-optic missile seekers and laser-guided bombs first proved their utility. Later, the Soviet-Afghan war again underscored the usefulness of night vision and low-cost electro-optic seekers on low-cost missiles. By the time of Desert Storm, warfare was largely fought and won at night, placing the priority for electro-optics as high as more traditional technologies such as radar and communications. Even early-warning satellites were used to detect incoming SCUD missiles. While all these wars were going on, interplanetary space probes, orbital platforms, and nuclear testing further augmented the development of EO systems, as film was not suitable for these applications. Today, thanks to television, wars, and space probes, the table has turned on film, and EO sensors are almost always the system of choice. The movie industry is rapidly moving from "filming" a movie to videotaping or computing the entertainment feature. Still photography is the final bastion for EO systems, and the advent of high-resolution CCDs will likely banish film to an obscure art form early in the next century.

The system designer is challenged on many fronts. The role is to evaluate the needs of a customer and derive the engineering requirements that are a satisfactory balance of four key topics:

- performance
- risk
- cost
- schedule

This usually starts with a detailed assessment of the performance requirements, conversion of them to concepts, and the eventual elimination of the weaker concepts. Throughout the process, consideration must be given to these four keys.

Once a concept or perhaps a few concepts have been selected, the system engineer can consider developing the performance requirements for each of the subsystems and the inclusion of design experts in those fields. A modern approach to this process, the so-called *concurrent engineering* approach, has the engineers from all appropriate disciplines (e.g., design, quality, field support), as well as installation and test engineers, participate from the very beginning so that they can ensure that the concepts under consideration are implementable. The system engineer has the responsibility for explaining the concept in enough detail that the detailed design process can be undertaken. General design rules, often in the form of rules of thumb, are used to judge the effectiveness of the concept before the design process can begin. The system engineer needs the skills to communicate with all of the members of the design team and must be able to facilitate the communication between the team members and know when communication is urgent. In particular fields, it can be very important to develop rules of thumb that all designers can understand. That way, the specialists in various fields can anticipate, with reasonable accuracy, the performance characteristics that will evolve in other design areas. For example, the controls designers should have an easy way to estimate the mass of the motors used to move some part of the system that is controlled by his part of the design. That way, even though he is not an expert on actuators and motors, he can know if his design meets the mass budget that has been allocated by the systems engineer. Clearly, this book does not address all of the rules of thumb that might come up in designs of systems, since they are so widely varied. However, the types of rules presented here can act as a guide for the rule development process and the eventual creation of general guidelines upon which all designers can rely.

The interested reader can find any number of texts that provide details on how various EO systems function. For example, Hecht has an entire section that describes telescopes, microscopes, and related systems. Of course, the reader who is interested in a specific topic will have to resort to books and technical journals dedicated to systems that apply in those cases. Often, EO systems are described in *Optical Engineering, OE Reports,* and SPIE and IRIA/IRIS conference proceedings. Additionally, the journal *Applied Optics* presents a description of EO systems in each issue, at a fairly brief and sophisticated level. Specific chapters and sections relating to electro-optical system engineering can be found in Hudson's *Infrared Systems Engineering,* Seyrafi's *Electro Optical System Analysis,* Spiro and Schlessinger's *Infrared Technology Fundamentals.,* Miller's *Principles of Infrared Technology,* and Holst's *Electro-Optical Imaging System Performance.*

USEFUL MAGNIFICATION OF A MICROSCOPE

Subject: Systems, microscopes

The Rule

$$\text{Necessary magnification} > \frac{\text{Visual acuity limit} \times D_v}{\dfrac{0.61\lambda}{NA}}$$

where D_v = normal distance of distinct vision which is about 10 inches
   λ = wavelength of light
   NA = numerical aperture

The visual acuity limit is one minute of arc, or two minutes for comfortable long-duration viewing. We note that the eye can resolve about 40 arcsec but clearly cannot maintain that performance for long periods of time.

Basis for the Rule

The rule is based on optical engineering and ergonomics.

Cautions and Useful Range of the Rule

The rule is applicable to microscope magnification for humans.

Usefulness of the Rule

It is useful for optical design and for determining useful eyepiece for a microscope.

Notes and Explanation

The closest object interval just resolved by a microscope must at least subtend an angle at the eye equal to one minute of an arc, or two minutes of an arc for more comfortable viewing. A value that supports this limit is known as *useful magnification,* and a magnification that exceeds this is known as *empty magnification.*

   The above principles can determine the power of the eyepiece required to go with any objective to accommodate the resolving power.

Source

B. K. Johnson. 1960. *Optics and Optical Instruments.* New York: Dover Publications, 93.

TYPICAL VALUES FOR EO SYSTEMS

Subject: Systems, radiometry, system performance

The Rule

When the details of a particular electro-optic system are unknown, Seyrafi[1] gives the following typical values that can be assumed for sensitivity calculations:

- overall transmission through the optics: 0.6
- electrical efficiency: 0.8
- scan efficiency: 0.8
- frame time: $1/30$ s for imaging systems
- atmospheric transmission: 0.7 in a transparent region

The authors of this book also suggest the following additional parameters:

- typical reflectivity of targets not intended to be "stealthy": 40 percent
- temperature of objects in space: 270 K
- detector D*: 10^{11} Jones (a Jones has the units of cm $\sqrt{\text{Hz}}$ /watt
- temperature of objects on the ground: 300 K
- optical cross section: $(0.1\ D^4)/\lambda^2$ m^2/sr (D = diameter in meters, and λ = wavelength in meters)

Basis for the Rule

Empirical approximations are based on hardware experience. Even though those rules contributed by Seyrafi date from 1973, they still generally apply today.

Cautions and Useful Range of the Rule

These are typical values for systems; however, any given system may have drastically different values. Use these only when no other information is available. Rarely will any of the above be off by more than a factor of ten. However, if more than one of these numbers is used, the actual result may be off considerably, and care must be taken to ensure that the end result is correct.

Usefulness of the Rule

The rule allows first-guess calculations when design details are unknown. This set of parameters also allows the designer to begin the design process and set up the design equations before the actual details are revealed.

These guidelines usually add sufficient margin so that the hardware can actually be made to perform to expectations.

Notes and Explanation

Often, one needs to perform quick calculations of sensitivity or determine the range of some system attribute (e.g., how big does the aperture need to be?) when incomplete design information is known. Although this can lead to dangerous design decisions, when necessary, the above can be substituted into the equations.

Reference

1. K. Seyrafi. 1973. *Electro-Optical Systems Analysis.* Los Angeles: Electro-Optical Research Company, 238, 294.

ELECTRICAL FREQUENCY BANDPASS FOR A SPECTROMETER

Subject: Systems, spectrometers, scanned signal rise times

The Rule

The electrical frequency bandpass for a spectrometer should be:

$$f > \text{ or } \approx 0.8RS$$

where f = electrical frequency bandpass
    R = resolving power of the spectrometer which is given by the wavelength divided by the incremental wavelength bandpass ($\lambda/\Delta\lambda$)
    S = scan rate in scans per second

Basis for the Rule

The electrical frequency bandpass must be $0.8RS$ so that the detection circuit can keep up with the changes in detector output. The ">" is included to indicate that enough bandwidth is needed to catch nuances in the spectrum.

The rule is based on spectrometer design state of the art and the rise time of a signal on an RC circuit.

Cautions and Useful Range of the Rule

The rule assumes that the integration time is five times the system time constant. In addition, this rule is true only if the resolving power is defined as the number of spectral lines per scan.

Usefulness of the Rule

The rule is useful for estimating electrical frequency bandpass for a spectrometer.

Notes and Explanation

As a sensor scans a scene, the input is approximately a square wave (actually, more rounded due to the optics). The sensor response is a exponential function related to the rise time. When the integration time is equal to 2.2 times the rise time, the hardware output achieves 80 percent of the theoretical (square wave) amplitude of the source. When the integration time is 5 times the dwell or integration time, the output achieves 99.3 percent of the amplitude and represents a nominal design value for accurate systems.

The electrical frequency response is related to the time constant by a simple RC network. The integration time is a function of the resolving power and the scan rate. All this can be combined to yield a frequency response equal to $5RS/2\pi$, which can be simplified to the above.

In general, design practices dictate that the time constant be equal to or less than the dwell time divided by five. This results in a loss of instrument sensitivity, but the time constant can be increased after the measurement by postprocessing. However, it never can be decreased.

Source

C. Wyatt. 1978. *Radiometric Calibration: Theory and Methods.* Orlando, FL: Academic Press, 145–148

System Off-Axis Rejection

Subject: Systems, bright source rejection

The Rule

It is difficult, but not impossible, to create sensors that can perform (and in some cases, not be damaged) when the sun on moon is within about 10° of their line of sight or when sun light is falling on the any of the optical surfaces, even if it is not within the field of view.

Basis for the Rule

A review of the stray light rejection of various telescope designs, such as are shown in *The Infrared Handbook*,[1] shows that the point source transmittance of a typical telescope is about 10^{-3} at 10°. While this is not true of all designs, it is a typical value. Since the irradiance at a sensor from even a small part of the sun or moon is higher than that of most stars and all man-made objects, rejections of 10^{-5} to 10^{-6} may be required. Such rejections are only easily achieved at off-axis angles of 30° or greater.

Chaisson[2] reports that highly sensitive astronomical telescopes, such as Hubble, cannot have the sun within 50° of the line of sight or the moon within 15°.

Cautions and Useful Range of the Rule

The exact optical design and radiometric system parameters will determine the performance of the system. These guidelines are a good first estimate of what can be expected, however. We do note, on the other hand, that there are some important exceptions. For example, laser warning and some forward-looking infrared sensors can function with the sun in the field of view.

Usefulness of the Rule

This rule allows system designers to make a first estimate of the exclusion angles required for missions in which the sun might be present.

Notes and Explanation

A critical part of the optical performance of a telescope is the cleanliness of the optical surfaces. Even the smallest surface contamination will cause light from out of the field of view to be scattered in a way that today's modern and very sensitive detectors will see. Therefore, while sensors are now more capable of sensing subtle features of the things they are pointed at, they also see undesirable light sources.

Reference

1. W. Wolfe and G. Zissis. 1978. *The Infrared Handbook*. ERIM, Ann Arbor, MI, 19-24, 19-25.
2. E. Chaisson. 1994. *The Hubble Wars*. HarperCollins, 72.

SYSTEM COMPLEXITY OF LONG WAVELENGTHS

Subject: Systems, sensor design

The Rule

Infrared sensor design complexity is proportional to sensor wavelength. For a given sensitivity (and/or resolution), the longer the wavelength, the more massive, power hungry, complex, and immense the sensor becomes.

Basis for the Rule

This rule is based on the state of the art of the technology, and a principle dictated by physics:

1. The aperture size must grow linearly with increasing wavelength to maintain resolution due to diffraction.

2. Some cooling usually becomes necessary as the wavelength is increased beyond the near infrared. This is particularly true as the wavelength is extended to include the infrared wavelengths beyond about 3 microns. Detector performance and the need for cooled optics at long wavelengths demand some level of cryogenic cooling. Generally, cooling the solid angle around the detector with a cold shield and the bandpass filter is required. The longer the wavelength, the colder it needs to be. This affects both the number of components to be cooled and their final temperature.

Cautions and Useful Range of the Rule

The rule is especially valid at wavelengths longer than visible. It does not apply in the UV or the submillimeter wave regime where longer wavelengths make things easier.

This rule does not completely apply to chopped pyroelectric cameras and other types of low-sensitivity systems that can ignore the thermal emission from the optics and do not require cooling. However, the diffraction effects still remain.

Modern advancements in uncooled LWIR arrays mitigate some of the effect, but the cooling requirements for optics often remain.

Usefulness of the Rule

The rule can be used to quickly illustrate the pain and difficulty of increasing a sensor' wavelength. This rule should motivate the designer to seek shorter wave bands whenever possible.

Notes and Explanation

It is always easier to design a given sensor with a shorter wavelength than one with a longer one, within the limits mentioned above. The shorter the wavelength, the less cooling and smaller aperture will be needed to achieve a particular level of performance in both sensitivity and angular resolution. Hence, the complexity, price, weight, and power consumption are reduced, and producibility and reliability are increased.

As the state of the art in cooling, lightweight optics, and FPAs progresses, the impact of this rule will decrease.

SPECTRUM PERSPECTIVE, OR THE DO NOT ENVY YOUR NEIGHBOR'S LAWN RULE

Subject: Systems, bandpass selection

The Rule

The spectrum is always greener on the other side of the bandpass (at least until you get there).

Basis for the Rule

The rule is based on empirical observations.

Cautions and Useful Range of the Rule

This is a gross generalization.

Usefulness of the Rule

The rule underscores the danger of assuming that all your problems can be solved by changing to another bandpass.

It also suggests that your competitors are also facing problems and would prefer to be in your bandpass.

Notes and Explanation

When a sensing need arises, there is generally a war between UV, visible, IR, and radar systems. For folks working in one spectral regime, it always seems that those working in the other have a technological advantage. This is rarely the case—usually, the phenomenology, background, and politics determine the most appropriate system rather than the technology.

One of the great equalizers in this type of decision making is the availability of all of the information needed to make informed decisions. For example, when designers compare the prospects of tracking targets with UV or infrared, they often are restricted to the IR solution because so little information exists on the spectral characteristics of typical targets in the UV. UV signatures are usually small and are highly scattered by the atmosphere.

SPACE SENSOR LARGE APERTURE PRODUCIBILITY

Subject: Systems, costs, sizes

The Rule

Space sensors with apertures up to 20 cm diameter are difficult and expensive to build, test, and qualify for space. Bigger apertures are nearly impossible in comparison.

Basis for the Rule

Any number of evaluations of space tracking systems have shown that large-aperture tracking systems are a nightmare in terms of cost and complexity. The reader should be cautious when reviewing results of studies, however, since they often stretch the truth (and frequently the laws of physics) in an effort to support the goals of the various funding agencies that were involved. The truth is that a system with about a 1 m aperture will cost about $1 billion dollars, but it is doable.

Of course, we're only joking about the 20 cm limit, but the cost and complexity of space-qualified optics is formidable and becomes even more critical at larger apertures. It is clear that linear scaling based on the diameter of the telescope is extremely naive. Other system solutions, such as reducing the range to the target, improving focal plane performance, choosing mission scenarios that reduce background radiation, and so on allow the designer to keep the telescope aperture small and are generally a preferable solution in any design.

Cautions and Useful Range of the Rule

Advancements in technology will bend this rule, but for now its not a bad approximation of the realities of space optics.

Usefulness of the Rule

This type of rule should be considered every time someone suggests that a larger aperture will solve the problem. The authors have seen this knee-jerk reaction applied to a number of systems, including optical communications, where the main penalty is the mass and cost growth when larger apertures are used.

Notes and Explanation

It should be no surprise that commercial telescopes stop at these modest sizes. At larger sizes, the combined effect of larger optical surfaces, the increased mass needed to provide stability, and the heftier mount needed to provide a solid base all combine to increase the cost and complexity of the telescope.

SOLID ANGLE OF SPECTROMETER

Subject: Systems, spectrometers, optics

The Rule

The solid angle field of view for a spectrometer can be approximated by $0.8\text{f}\#^{-2}$.

Basis for the Rule

The rule is based on approximations and simplification of geometric optics. In general, any optical system has

$$\omega \approx \frac{\pi}{4\,f\#^{2}}$$

which approximates the above rule.

Cautions and Useful Range of the Rule

The rule is useful for f numbers greater than 1.5.

Usefulness of the Rule

The rule is useful for estimating the solid angle of a spectrometer.

Notes and Explanation

Spectroscopists are normally interested in the radiation within a small solid angle normal to the receiving aperture. This solid angle depends on the angle that the optics can accept (field of view). This is crudely related to the focal length.

Source

M. Kimmitt. 1970. *Far-Infrared Techniques*. London: Pion, 44.

RESOLVING POWER OF A SPECTROMETER, OR JACQUINOT'S RESOLVING POWER

Subject: Systems, radiometry, spectrometers

The Rule

The resolving power of a spectrometer is approximated by:

$$R \approx 6\left(\frac{f}{d}\right)^2$$

where R = resolving power
$\quad f$ = collimator's focal length
$\quad d$ = source diameter

Basis for the Rule

The rule is based on approximations and simplification of diffraction and geometrical optics theories. Jacquinot derived this based on a source subtending a solid angle (ω) at the collimating mirror by

$$R = \left(\frac{\bar{\upsilon}}{\Delta\bar{\upsilon}}\right) = \left(\frac{2\pi}{\omega}\right) = 6\left(\frac{f}{d}\right)^2$$

Cautions and Useful Range of the Rule

Actual resolution may be limited by other factors such as apodisation, and this is an approximation only.

Usefulness of Rule

It is used for estimating the resolving power of a spectrometer.

Notes and Explanation

The resolving power of any interferometric spectrometer is limited by the beam spread of off-axis rays caused by imperfect collimation. This spread also leads to a small change in ray path and causes the wavenumbers of the computed spectrum to be underestimated by a small factor.

Kimmitt[1] gives the original reference for this work as P. Jacquinot. 1960. *Rept. Progr. Phys.* 23, 267.

Reference

1. M. Kimmitt. 1970. *Far-Infrared Techniques*. London: Pion, 105.

PROCEDURES TO REDUCE NARCISSUS EFFECTS

Subject: Systems, Narcissus effects

The Rules

Lloyd[1] has suggested the following five design procedures to reduce cold reflections (sometimes known as *Narcissus effects*), which cause the detector to see reflections of itself:

1. Reduce the focal plane effective radiating cold area by baffling.
2. Reduce lens surface reflections by using high-efficiency antireflection coatings (on both sides of the optical elements).
3. Defocus the potential cold return by designing the optical system so that no confocal surfaces are present.
4. Cant (or tilt) all flat windows. This means that rays traveling parallel to the line of sight of the detectors will be diverted out of the sensor line of sight.
5. When all else fails, null out the cold reflections with a thermal source or by electronic image processing.

Cautions and Useful Range of the Rule

On any given design, requirements and trade-offs rarely allow all five to be implemented.

Usefulness of the Rule

These are good practices when designing an IR system.

Notes and Explanation

Often, with IR system design, reflections from cold surfaces onto the FPA cause a display irritation of low level in those areas. Lloyd's five techniques serve to reduce this unwanted effect.

Sometimes method no. 1 cannot be done. For LWIR systems, no. 2 should always be done if budget and throughput requirements allow. Number 3 should be done whenever the optical requirements allow. Numbers 4 and 5 are simple yet effective way of reducing this effect; they should be employed whenever possible.

Reference

1. J. Lloyd. 1975. *Thermal Imaging Systems.* New York: Plenum Press, 281.

GOOD FRINGE VISIBILITY

Subject: Optics, interferometers, shop optics

The Rule

Good fringe visibility from a disk source occurs when

$$\frac{\pi h \theta}{\lambda_o} = 1$$

where h = distance between the two slits (or mirrors) forming the fringes
   θ = angular separation between two point sources
   λ_o = wavelength of a narrow bandwidth

Basis for the Rule

This rule is based on the Van Cittert-Zernike theorem. It has been verified by empirical confirmations. The fringe visibility is related to the degree of coherence of the optical field at each mirror of the interferometer and can be modeled as a Bessel function.

Cautions and Useful Range of the Rule

This is valid for narrow spectral bands only. It assumes a disk source of nearly constant intensity and assumes that good coherence exists and corresponds to a "visibility" of 0.88.

Usefulness of the Rule

This rule is useful for determining when a source will produce visible fringes such as required to measure the angular size of a disk of some object (e.g., a star). This is useful for experimental design and in the design of a Michelson stellar interferometer. It is also useful for interferometer design considerations.

Notes and Explanation

For a narrowband source of wavelength λ_o and diameter D at a distance of R, there is an area of coherence $\pi(h/2)^2$. The angular size of the object $\theta = (D/R)$, and the transverse correlation distance is $0.32(R\lambda_o/D)$. A set of apertures at least this close together will produce highly visible fringes.

Source

E. Hecht. 1987. *Optics.* Reading, MA: Addison-Wesley, 437.

GIMBAL TO SLEWED MASS

Subject: Systems, gimbals, servos, control loops

The Rule

The mass of a gimbal assembly is directly proportional (or raised to a slight power such as 1.2) to the amount of slewed mass.

Basis for the Rule

The rule is based on typical industry experience for state-of-the-art systems.

Cautions and Useful Ranges of the Rule

To use this scaling, accelerations, velocities, base motion rejection, and stability must be identical; slew masses should be within a factor of 3 of each other; and the gimbals must be of the same type and material and of same number of axes. Finally, environmental, stability, stiffness, and slewing specifications must be similar.

Usefulness of the Rule

The rule illustrates system trade-offs by providing data on the mass impact of changing the slewed sensor mass (e.g., optics size). It is also useful for estimating if a given gimbal requires advanced technology to meet mass goals.

Notes and Explanation

A given gimbal's mass depends to a great extent on its design, composition materials, size, required accelerations, base motion rejection, stability, the light weighting techniques applied, and the mass to be pointed. An estimation of the mass of an unknown gimbal assembly can be scaled, based on the known mass of a similar gimbal. Past designs and hardware indicate that the gimbal mass scales approximately linearly to the mass to be slewed.

EXPECTED MTF RULE

Subject: Systems, MTF, figure of merit

The Rule

For typical sensors and a good display, the diffraction and the detector's modulation transfer function (MTF) will dominate the system's MTF. Diffraction MTF is typically 0.7, detector MTF is 0.64, design MTF blur is 0.9, and processor MTF is close to 1. The sensor MTF is then about 0.4, because the combined MTF is obtained by a root sum of squares process. Displays are difficult to generalize, but most achieve at least 0.6. Thus, one can expect the system MTF to be in the range of 0.25 to 0.4.

Basis for the Rule

The rule is based on empirical observations. It is also important to note that these values are given at f_o, which is defined as $1/(2IFOV)$, where $IFOV$ is the instantaneous field of view.

Cautions and Useful Range of the Rule

The suitability of the rule depends on the design.

Usefulness of the Rule

It is useful for estimating component and system MTFs.

Notes and Explanation

The total MTF of a system is the product of the MTFs of all of the subsystems.

Source

G. Hopper. 1993. Forward looking infrared systems. In *Passive Electro-Optical Systems*, vol. 5, ed. S. Campana, of *The Infrared and Electro-Optical Systems Handbook*, executive ed. J. Accetta and D. Shumaker. Ann Arbor, MI: ERIM, and Bellingham WA: SPIE, 128–130.

COOLING OPTICS

Subject: Systems

The Rule

Optics will not need to be cooled if the bandpass cutoff is ≈ 12 μm or less and if the baffle design and FOV of the FPA are properly designed. Generally, longer wavelength systems will require cold optics.

Basis for the Rule

The rule is based on empirical observations and the state of art of current detectors. In addition, the proper applications of the laws of cooled optics allows for the inclusion of re-imaged systems. In such systems, the hemisphere into which the detector looks is almost totally cooled, except for a small solid angle that includes the target.

Cautions and Useful Range of The Rule

Some readers will notice that forward-looking infrared systems usually do not have cooled mirrors but still operate effectively. The reason is that they generally are looking for targets that are relatively warm, and they frequently employ detectors that are noisy enough that the emissions from high-quality (low emissivity) mirrors are insignificant in comparison to other noise. Nevertheless, they are almost always cold shielded.

Excessively noise-free detectors or very narrow IFOV detectors will still benefit from cooled optics, even if their bandpass is shorter than above-defined cutoff.

The success of this rule depends on the spectral nature of the background and the emissivity of the optics that are employed. In any case, the structure and optics in the vicinity of the detector must be cooled for most applications. Nevertheless, cooling the optics usually results in increased performance (unless some other noise source is extremely dominant).

Usefulness of the Rule

This rule is effective when the bandwidth of the system is being chosen. For systems where the target is close to 300 K, some compromise in performance is implied, because most of the energy will be the vicinity of 10 microns, and cooling of both optics and FPA usually must be considered. If the targets are warmer, then the designer has some latitude and may be able to avoid the longer wavelengths that create problems in implementing the design.

Notes and Explanation

Thermal emission from the optics is a noise source for detectors. The amount of noise generated by the thermal emission of the optics is usually negligible for bandpasses with a cutoff of less than 11 microns or so. For bandpasses (with cutoff) of less than about 10 microns, there is little or no sensitivity to be gained by cooling the optics. For bandpasses including wavelengths of 12 microns or more, the optics will almost certainly need to be cooled.

BLIP LIMITING RULE

Subject: Systems, noise

The Rule

There is little to no benefit in having a sensor more sensitive than the noise imposed by the background.

Basis for the Rule

The rule is a derivation of the resultant system noise properties of sensors.

Cautions and Useful Range of the Rule

The noise from the background is much larger than other noise sources; that is, the noise and the sensor must be BLIP.

Usefulness of the Rule

It is useful in estimating the needed detector sensitivity.

Notes and Explanation

A significant noise source in the infrared is the noise caused by the inconstant arrival rate of the photons from the background. This fluctuation in the arrival rate of the photons is an unavoidable feature of the radiation source. For high backgrounds (large IFOVs, bandpasses, and integration times) this is frequently the driving noise source. Expending money and effort to get more sensitive detectors will not increase overall system sensitivity.

APERTURE VS. PACKAGE

Subject: Systems, gimbals

The Rule

It is difficult to package a gimbaled aperture in a volume where the ratio of aperture diameter to package diameter exceeds about 0.6. It is difficult to package a nongimbaled aperture in a volume where the ratio of aperture to package diameter is 0.8 or more.

Basis for the Rule

The rule is based on empirical observations for state-of-the-art gimbal systems in typical installations.

Cautions and Useful Range of the Rule

This rule does not apply to systems that do not have a packaging constraint, such as ground-based astronomical telescopes. This rule is a generalization, of course, since individual designs and requirements should be fully analyzed and traded. Optical systems may press the limits for these ratios, but the presence of cryogenic cooling lines, electrical cables, and other connections to the outside world unavoidably take up space.

Usefulness of the Rule

This rule makes a rough and quick estimate of the volume that must be set aside for a sensor.

Notes and Explanation

When attempting to package an electro-optical instrument, the ratio of the aperture to the total size tends to follow the above rule. Although there have been some systems that violate this rule, they required enormous levels of engineering ingenuity and complexity. This, of course, translates into added cost for the system. We usually find that additional impacts result, such as the need to place nearby structures in precarious positions or limit the operational range of the sensor.

17

Target Phenomenology

Generally, the properties of targets and their signatures, such as are summarized in this chapter, falls into the domain of the military infrared tracker designer. Increasingly, many of these rules apply to nonmilitary segments of the EO market such as paramilitary, search and rescue, environmental monitoring, general surveillance, remote diagnostics, and industrial security.

This chapter provides a brief look into the short-cut characterizations that were largely developed in the military tracking arena to assess the signatures of various engine plumes and hardbodies. Regardless of their heritage, several of these rules are applicable in the generic business of assessing what a sensor might see when tracking an appropriate target (unfortunately, most of these rules are limited to missiles or aircraft). Although most of these rules were developed for the infrared spectrum, they illustrate important principles that may be applied (albeit with caution) to other parts of the spectrum.

Often, targets of interest to the EO sensor designer consist of the metal body and frame containing some kind of engine (such as your car). Such a target can be detected by sensing the metal hardbody (e.g., the roof of your car) or the engine emission (e.g., the hot CO_2 coming out of your tailpipe). The emission of hot gases and particles is generally called a *plume*. Although all man-made engines produce significant plumes, those of jet engines and rockets draw the most attention.

Rocket and jet plumes have long been of interest to EO designers, as these provide bright signature-to-background ratios. Much early work was done in the 1950s and 1960s in remote plume diagnostics by electro-optical instruments for jet and rocket engine development. Interestingly, at least one major contemporary sensor company developed from a support group charged to develop sensors to characterize their company's rocket engines. Much effort was expended in the 1960s on large and small rocket plume signatures, thanks to the space and arms races. The signatures of tactical air-to-air and surface-to-air missiles were investigated in the hope of providing effective countermeasures. Plume investigations of large rockets continued in support of early warning efforts. Maturation of this study and the perceived need were formalized during the United State's Strategic Defense Initiative (SDI) era in which significant effort was expended in refining the plume and hard-body signatures of large missiles. This tradition is continuing due to recent prolifer-ation of missiles of all types, requiring defensive weapons to protect against these weapons, which in turn require accurate target characterization.

As expected, there has been considerable interest in the reflectivity of target sur-faces, because this determines the amount of laser tracker radiation that can be ex-pected to come back to a receiver, determines the visible-band signature, and affects the IR signature. Ironically, one of the most challenging features of charac-terization of a threat signature is determining the reflectivity of the plume at typical laser radar wavelengths. The most difficult of these is the liquid fueled rocket, since the emissions may consist primarily of water and other hot gases that appear nearly transparent, especially in the visible wavelengths. The Space Shuttle, for example, produces a large white plume from its solid rocket motors, but the plumes from the hydrogen-oxygen engines are nearly transparent. In fact, one can see directly into the engine housings (during launch) on the three LOX-LH engines built into the orbiter. Although much less intense than the infrared signature, the visible and UV signatures are of interest because of the availability of their hardware technologies and improvement in diffraction, and because the plume is more confined in spatial extent is located in a predictable place with respect to the rocket. All of these fac-tors facilitate accurate tracking.

By the 1970s, computing power became sufficient to spur various governments and companies toward developing computer codes that allow system engineers to estimate the magnitude of the signatures in a given spectral bandpass. Some of the most frequently used ones include the JANNAF (Joint Army, Navy, NASA, Air Force), PLURAD (plume radiance code), SPF-2 (the Standard Plume Flow Field) and SIRRIM (Standardized IR Radiation Model) codes that were geared to estimate signatures from tactical missiles based on complex chemistry and flow fields. Per-haps the most complex and costly plume code is CHARM (Composite High Alti-tude Radiance Model), a model intended to estimate large pocket plume signatures and shape by assessing the chemistry of the rocket fuels used and their interaction with the rarefied atmosphere. The signature code craze is not limited to the plumes, as complicated codes have been developed to predict laser cross section, reflectivity, and hardbody signatures throughout the spectrum [e.g., the Spectral Infrared Thermal Signatures (SPIRITS) and the Optical Signature Code (OSC)]. Unfortunately, such codes occasionally proved to be somewhat less than perfect in actual applications, providing accuracies no better than about one order of magni-tude.

For the reader interested in more details, the *Infrared-Electro-Optical Handbook* is about the best resource in the open literature. It contains information on almost ev-ery topic related to targets and backgrounds. Another source that should not be overlooked are the older versions (similar compilations) variously called *The Infra-red Handbook* and variations of that name. Look for the older versions in used book stores and in the offices of older employees. Many specific signature handbooks ex-ist from governments and corporations active in the field, and these can provide valuable insight.

One can find introductory discussions of likely scientific, commercial, tactical, and strategic targets in Hudson's *Infrared System Engineering* and Miller's *Principles of*

Infrared Technology. Likewise, Spiro and Schlessinger's *Infrared Technology Fundamentals* has significant discussions of likely strategic targets, and Holst's *Electro-Optical Imaging System Performance* has discussions of tactical military targets. For up-to-date detail measurements, explanations of phenomena, and code development do not overlook the frequent publications of the SPIE, the JANNAF working groups, and the IRIA/IRIS conferences and proceedings.

TEMPERATURE AS A FUNCTION OF AERODYNAMIC HEATING

Subject: Target phenomenology, aerodynamic heating, systems

The Rule

The maximum temperature caused by aerodynamic heating is[1]

$$T = T_{amb}\left[1 + r\frac{(\gamma - 1)}{2}M^2\right]$$

where T = stagnation temperature in Kelvins

T_{amb} = ambient temperature

r = recovery factor (usually 0.8 and 0.9) (For laminar flow, use r = 0.85, and for turbulent flow, use r = 0.89)[2]

γ = ratio of specific heats of air at 25° C (usually 1.4)

M = mach number

With some assumptions, the first equation can be further simplified for high-altitude flight as follows:

$$T = 217(1 + 0.164M^2)$$

Basis for the Rule

This rule is based on aerodynamic theory and empirical measurements.

Cautions and Useful Range of the Rule

This really represents the stagnation temperature of the air at the surface of the object when moving directly against the air. The actual temperature of the object will be lower. For instance, the temperature of a dome will fall off rapidly as the distance from the central part increases. This is accounted for by r, the recovery factor. Hudson suggests using 0.82 for laminar flow and 0.87 for turbulent flow.[1]

Usefulness of the Rule

This rule is interesting in that it can be used to estimate the temperature of a target It can also be used to estimate the temperatures that will be encountered in designing sensor components, such as windows, that will be used in various types of airborne sensors. This provides a basic piece of information about the design of such devices because elevated window (or dome) temperatures may result in noise, which may need to be considered. The temperature and the emissivity of the window material determine how much background radiation flux (and therefore noise) the window adds to the sensor.

Notes and Explanation

The first equation applies for mach numbers less than about 6. For higher mach numbers, particularly above 8, the T_{amb} term should be divided by about 2.[3]

References

1. R. D. Hudson, Jr. 1969. *Infrared Systems Engineering* New York. John Wiley & Sons, 101.

2. J. Accetta. 1993. Infrared search and track systems. In *Passive Electro-Optical Systems,* vol. 5, ed. S. Campana, of *The Infrared and Electro-Optical Systems Handbook,* executive ed. J. Accetta and D. Shumaker. Ann Arbor, MI: ERIM, and Bellingham WA: SPIE, 223.

3. K. Gilbert et al. 1993. Aerodynamic effects. In *Atmospheric Propagation of Radiation,* vol. 2, ed. C. Fox, of *The Infrared and Electro-Optical Systems Handbook,* executive ed. J. Accetta and D. Shumaker, Ann Arbor, MI: ERIM, and Bellingham, WA: SPIE, 241.

ROCKET PLUME RULES

Subject: Target phenomenology, plumes

The Rules

1. The size of the plume increases in diameter with altitude.

2. The intrinsic shock structures expand greatly with altitude, eventually enveloping the entire rocket at highest altitudes, just before burnout.

3. A minimum in infrared intensity is observed near the time that the missile velocity and exhaust velocity are the same in magnitude. This minimum is generally observed at missile altitudes from 70 to 90 km—in many cases coincidentally close to the time of staging.

Basis for the Rules

The rule is based on rocketry and empirical observations.

Cautions and Useful Range of the Rules

A number of phenomena cause these rules to be approximations only. For example, the *high-altitude trough* causes the signature of a rocket to get brighter with altitude but then diminish for a range of higher altitudes. This is the result of reduced afterburning of fuels that are not consumed in the engine, which occurs at lower altitude. The reduced afterburning results from the lower oxygen content in the thinner atmosphere. As the rocket continues to accelerate, brightness returns to the plume because the unburned fuel, although in a low-oxygen environment, again encounters enough oxygen to stimulate burning because of the rocket's increased speed.

Usefulness of the Rules

These rules are useful in predicting the signatures that will be available for tracking during rocket flight. They allow estimation of the performance of sensors by providing the necessary target spectral characteristics.

Notes and Explanation

These phenomena occur because of the diminishing effects of the atmosphere as the rocket ascends, and the rules are based on empirical observations of current rocketry.

Source

I. Spiro and M. Schlessinger. 1989. *Infrared Technology Fundamentals*. New York: Marcel Dekker, 60–62.

REFLECTIVITY OF METALS, OR THE HAGAN-RUBENS RULE

Subject: Target phenomenology, optics

The Rule

The reflectivity of metals can be estimated by:

$$R \approx 100 - \frac{3.7\,\rho^{0.5}}{\lambda^{0.5}}$$

where R = reflectivity (in percent)

ρ = resistivity of the metal in $\mu\Omega/m$ (e.g., for copper, this is $0.02\,\mu\Omega/m$)

λ = wavelength (in microns)

Basis for the Rule

The rule is based on empirical observation and theory.

Cautions and Useful Range of the Rule

This rule applies to the wavelengths from the visible through IR. It describes the properties of smooth surfaces only.

Usefulness of the Rule

This rule can be used for a first-cut quick estimate of reflectivity. It also provides an estimate of the way the reflectivity of materials changes with wavelength. This can be useful in estimating the target signature from reflected sunlight or laser illumination, if the reflectivity is known at one wavelength.

Notes and Explanation

With some substitution and reasonable assumptions, the rule can also be expressed as

$$R = 100 - 100\frac{2}{c}\sqrt{\frac{4\pi\upsilon}{\mu\sigma}}$$

where R = reflectivity (in percent notation)

υ = frequency of the light radiation

σ = electrical conductivity

c = speed of light

μ = absolute magnetic permeability

The amount of reflection of light from a given (solid) object depends on the exterior surface morphology and surface coating. The above rule addresses smooth surfaces only. The reflectivity of a metal is related to the index of refraction of the surface coating. This can be estimated for a given wavelength by the metal's absolute magnetic permeability and electrical conductivity.

Source

D. Fisher. 1988. *Rules of Thumb for Scientists and Engineers.* Houston, TX: Gulf Publishing.

PLUME THRUST SCALING

Subject: Target phenomenology, rocket plumes

The Rule

One can scale the signature of one missile's plume to that of another by the size of the missile's thrust, or

$$I_1 = I_2 \left(\frac{N_1}{N_2} \right)^x$$

where I_1 = in-band radiant intensity of plume 1

   I_2 = in-band missile radiant intensity of Plume 2

   x = a constant depending on spectral bandpass [The constant, x, is usually between 0.7 and 2; assume 1 (linear) if you do not have specific data for the missile type and fuel.]

   $N_{1,2}$ = thrust in newtons for missiles 1 and 2

Basis for the Rule

The rule is based on empirical observations.

Cautions and Useful Range of the Rule

Do not scale across different spectral bands, as missile plumes are strong spectral-selective emitters. A slight change in bandpass prevents accurate scaling.

   Also, scale only similar fuels, motors, and motor geometries, and only for the same altitudes.

Usefulness of the Rule

The rule allows for easy estimation of the signature of a rocket, if the signature of a similar rocket of another size is known.

Notes and Explanation

Additionally, Wilmot[1] gives a scaling law for the viewing angle variations in observed signatures.

$$I_\theta = I_{90} \sin (\theta + \phi)$$

where I_θ = radiant intensity of the missile when observed at angle θ, with 90° representing a view from the side

   I_{90} = intensity at the beam viewing angle of 90°

   θ = angle between the velocity vector (frequently the same as the axis of the plume) and the observer

   ϕ = offset angle, a small correction whose value depends on the geometry of the missile and plume [This may compensate for the difference between the velocity vector and the plume axis (if not aligned).]

Signatures from small tactical missiles are typically from 100 to 10,000 W/sr/μm, depending on bandpass and aspect angle. ICBM and payload-orbiting rockets typically range from 10^5 to 10^7 W/sr/μm.[1] The in-band radiant intensity of a missile is proportional to the rate of fuel combustion, and that is proportional to the thrust of the motor. Therefore, signature (within a defined band) tends to scale almost linearly in relation to thrust.

Reference

1. D. Wilmot et al. 1993. Warning systems. In *Countermeasure Systems*, vol. 7, ed. D. Pollock, of *The Infrared and Electro-Optical Systems Handbook*, executive ed. J. Accetta and D. Shumaker, Ann Arbor, MI: ERIM, and Bellingham, WA: SPIE, 19–21.

PLUME DIAMETER

Subject: Target phenomenology

The Rule

A plume's brightness in just about any band varies approximately linearly with the thrust. Additionally the diameter of the plume is

$$D \approx \frac{\sqrt{F}}{\sqrt{P}}$$

where D = diameter of the plume
      F = thrust in units that agree with P
      P = ambient pressure in units that agree with F

Basis for the Rule

This is based on approximation of flow-field calculations and inspired observations.

Cautions and Useful Range of the Rule

The rule was devised for rockets, but it may be applied to jets, small missiles, cars and other vehicles with caution.

It assumes that the exhaust velocity greatly exceeds vehicle velocity, which is not the case when the rocket has been thrusting for some time and is moving very fast.

Usefulness of the Rule

The rule is useful for scaling a plume's diameter and radiant intensity.

Notes and Explanation

The plume from a rocket or jet usually (depending on the bandpass and aspect viewing angle) contains most of the usable signature that will be seen by a tracker. Within a band, for a given resolution and at the same altitude, the signature varies (can be scaled) as the thrust is increased, as described in another rule in this chapter. The signature varies in a complex fashion but, when all is said and done, it is usually is pretty close to linear.

The diameter of the plume can be estimated based on momentum balancing the plume pressure and the effective atmospheric pressure. Most of the signature will be contained within this diameter.

JET PLUME PHENOMENOLOGY RULES

Subject: Target phenomenology, plumes

The Rules

1. The radiance ($W/m^2/sr$) of a plume from a jet aircraft engine at 35,000 ft is about half the radiance at sea level.

2. It is better to observe a plane's plume in the CO_2 band at 4.3 microns than the water band at 2.7 microns.

3. The extent of a plume from a plane is roughly equal to the length of the plane.

4. A turbojet engine can be considered a graybody with an emissivity of 0.9, a temperature equal to the exit gas temperature, and an area equal to that of the exhaust nozzle.

5. For a subsonic aircraft, the exhaust temperature after expansion (i.e., the exhaust plume gas temperature) is approximately equal to 0.85 multiplied by the exhaust gas temperature (in the tailpipe) in Kelvins.

Basis for the Rules

These rules are based on astute empirical observations over a number of types of aircraft engines.

Cautions and Useful Range of the Rules

These rules assume normal aircraft operation; that is, the engine is not set for "afterburning."

It assumes that the expansion is to ambient pressure.

It assumes a classical turbojet engine design.

These assertions are very bandpass dependent.

Usefulness of the Rules

This is useful for estimating jet plume signatures to support the design of optical and infrared trackers.

Notes and Explanation

A jet engine burns fuel with the atmosphere's oxygen and produces a rocket-like thrust. Generally, the radiance from a turbofan is less than that from a turbojet. Also, in LWIR, the heat of the jet engine cavity has a much larger signature than the plume.

Source

R. D. Hudson, Jr. 1969. *Infrared Systems Engineering.* New York: John Wiley & Sons, 86–90.

FIRE PHENOMENOLOGY RULE

Subject: Target phenomenology

The Rule

Fires tend to flicker at 12 Hz and produce a spectrum of a 2000 K graybody of emissivity 0.7.

Basis for the Rule

The rule is based on empirical observations and measurements.

Cautions and Useful Range of the Rule

The actual characteristics of a particular flame or fire are highly dependent on the material being burned, but this rule applies for most common sources of flame.

This rule is based on the visible spectrum of a coal fire.

Usefulness of the Rule

This rule is useful in estimating target signatures associated with open flames.

Notes and Explanation

Electro-optical fire detectors often exploit the 12 Hz average flicker as a fire indication. Perhaps this 12 Hz flicker is tied to human evolution and the irritation at viewing displays that flicker near or at 12 Hz.

EMISSIVITY APPROXIMATIONS

Subject: Target phenomenology, emissivities

The Rule

If you do not know the emissivity of an object, assume 0.3 for metals, 0.8 for anything else, except 0.9 for anything that is claimed to be "black."

Basis for the Rule

The rule is based on empirical observations and a quick survey of material properties from various optical references.

Cautions and Useful Range of the Rule

It is applicable to approximations only.

Usefulness of the Rule

It provides a quick estimate of an object's emissivity when little is known about the object.

Notes and Explanation

It is important to know an object's emissivity when doing calculations related to target signatures. The emissivity can cause an order of magnitude change in irradiance. Generally, metals are of low emissivity, with optically smooth, polished surfaces of 0.01 or less. For most metal objects, the emissivity is closer to 0.3. It tends to be 0.8 for most other objects. In addition, if the sensor is looking into a cavity, such as a jet or rocket engine, the emissivity will tend to approach unity, regardless of the engine material.

EFFECTS OF SUNLIGHT

Subject: Target phenomenology, solar illumination

The Rule

Sunlit objects are often detectable to very long distances when viewed against a space background. The unfortunate circumstance is that debris and space trash illuminated by the sun may be the most detectable stuff in the field of view. Furthermore, space dust and such that is not in the field of view will be a source of decreased contrast within the field of view, which is more bad news.

Basis for the Rule

These types of problems have been regularly encountered in space sensors. Careful preparation of the space system to reduce the amount of debris it generates is critical to controlling this problem.

Cautions and Useful Range of the Rule

This is a general rule and must be considered in every space system that employs target sensing in the presence of solar radiation. This rule is also generally applicable to any sensor viewing against a low background.

Usefulness of the Rule

This rule reminds us to be cautious with the design and flight preparation of space sensors.

Notes and Explanation

The higher the sensitivity, the longer the sensing ranges (neglecting clutter and background). However, the higher the sensitivity, the more debris and unwanted, out-of-focus, trash will be detected. Therefore, a sensitive sensor operating with a low background may still be clutter limited! If one gains signal to noise by making the sensor more sensitive, then false alarms reach a level where they level off or even increase due to the sensor detecting specks of dust, debris, and so forth. For instance, a sunlit paint chip a few meters in front of the aperture of a sensitive, low-background space sensor will be the brightest thing in the field. In short, no matter what you are viewing, the more sensitive your sensor is, the more susceptible the sensor is to the presence of nearby debris.

If you operate the sensor in the presence of sunlight, you must include baffles and sun shades. These must be large if the they are to be effective, especially if the field of regard is large. Even when effective shades and baffles are employed, small items in the field of view will appear as targets. Focal plane processing may be needed to reject these items by assessing their angular velocity and acceleration.

Causes of Pigment's Color

Subject: Target phenomenology, pigments

The Rule

An appearance of white can be achieved with small pieces of transparent or translucent material as long as they are large enough in comparison with the wavelength of light, to allow for multiple refractions and reflections. This is the case with sugar, snow, beaten egg whites, clouds, and so on.

Basis for the Rule

This is a general observation from a variety of examples and is widely used in creating white appearances.

Cautions and Useful Range of the Rule

Since the appearance of "whiteness" depends on matching the size of the scattering material to the wavelength of light, paints of this type do not work over wide wavelength ranges. Infrared reflectivity may be considerably different from that in the visible regime. In addition, one must be certain that the materials added to the clear matrix have a different refractive index; otherwise, the material will look uniformly clear.

Usefulness of the Rule

This rule offers an easy approach for creating high-reflectivity surfaces without resorting to exotic approaches such as retro-reflectors.

Notes and Explanation

It is shown in a number of texts that scatter from particles depends on the relation of sizes of particles and the wavelength of light. Rayleigh showed that for particles about one-tenth of the wavelength of the light, scattering goes as $(1/\lambda^4)$, and for larger particles the scattering goes as $(1/\lambda^2)$. This is the root of the famous "blue sky" question that Rayleigh answered by showing that the scattering occurs in molecules that are small with respect to the wavelength of light. Shorter wavelengths are scattered more effectively; hence, we see scattered sunlight predominantly in the blue part of the spectrum.

One of the authors (Friedman) has also had the experience of using powdered quartz as a means of inducing turbidity into very clear and clean waters. The quartz was added to provide a measurable and repeatable turbidity.

Source

E. Hecht. *Optics.* 1990. Reading, MA. Addison-Wesley, 114.

Glossary

This glossary has been developed specifically to aid the reader of this volume in interpreting unfamiliar terms. The definitions are intended to have a practical, rather than formal, presentation. That is, they may be specifically oriented toward the use of the word in the electro-optics sciences, rather than a more general definition that would be found in a generic encyclopedia or dictionary.

aberrated
: Defines a range of degradations in image quality. Aberrations include, but are not limited to, field curvature, distortion, coma, chromatic aberration, and astigmatism.

absorptance
: The property of a material that acts to reduce the amount of radiation traversing through a section of the material. It is measured as the fraction of radiation absorbed traversing the material. Generally, the bulk absorptance of radiation by a material follows Beer's law.

adaptive optics
: Optical subsystems with the capability to change the wavefront in real time. They are generally used to compensate for atmospheric turbulence.

afocal
: Describes an optical system not forming a focus, such as often employed in astronomical telescopes and in laser beam expanders. Such optics accept a light field and produce an unfocused beam.

airmass
: A term frequently used by astronomers, referring to the path length in the atmosphere through which the telescope is looking. When a telescope is pointed straight up, it is looking through an airmass of one; the airmass increases as the telescope is pointed toward the horizon, as a function of the secant of the zenith angle.

Airy disk
: The distribution of radiation at the ideal focus of a circular aperture in an optical system when dominated by diffraction effects (e.g. no aberrations). The central spot formed by about 84 percent of the total energy.

albedo
The current use of the term is to indicate the total reflection and scatter from a surface, measured over all possible scattering angles and over some part of the solar spectrum which, if not specified, includes all wavelengths. Historically, astronomers have used *albedo* to indicate reflection in the visible wavelengths only, not all wavelengths.

algal
An adjective describing the properties of unicellular plants that can have a significant impact on the optics of the ocean.

Angstrom
A unit of length, equal to 10^{-10} meters, that is used to define wavelength, although it has been generally replaced by microns or nanometers.

antireflection coating
A coating intended to reduce the amount of radiation reflected from surface of an optical element.

apodisation
The modification of the transmission properties of an aperture to suppress unwanted optical aberrations or effects.

apostilb
A unit of measure of light in the system that takes account of the response of the human eye. An apostilb is equal to 1 lumen/m^2 for a perfectly diffuse (Lambertian) surface.

arcsecond
A measure of angle, usually employed by the astronomical community. An arcsecond is about 4.6 microradians and is computed as $1/60$ of an arc minute (which is $1/3600$ of a circle). Thus, an arcsecond is $1/60$ of $1/3600$ of 2π radians.

ASIC
Application-specific integrated circuit. These are custom-made chips "hardwired" to do specific functions. They provide the lightest weight and lowest-power processing but usually have little or no reprogramming ability.

aspheric
Not having a spherical shape. These shapes are often used in optics to define lenses and mirrors that are employed in advanced optical systems.

ATP
Acquisition, tracking, and pointing. A field of study focused on sampling theory, control theory, and servo systems. This is especially critical in search and rescue and weapon systems. Conversely, this usually is not a concern for navigation sensors or general cameras. ATP systems are sometimes employed to detect targets, control the line of sight, follow the motions of the target, and aim the device (e.g. another sensor, laser, or weapon) at the target. The latter must take into account the speed of the device aimed at the target.

avalanche photodiode
A semiconductor detector device that exploits a phenomena of photoemission and self-amplification.

azimuthal
When used as part of the coordinate system of a pointing or tracking system, for ground-based systems such as astronomical telescopes, azimuth usually refers (to rotation) to the plane that is parallel to the earth. For many systems (e.g., aircraft, spacecraft), the definition is often somewhat arbitrary but,

once chosen, it must be used by all members of the design team. Azimuth is the compliment of elevation, usually orthogonal to elevation in pointing coordinates.

background   The part of a sensed scene that is not the item of interest (e.g., the ocean is the background for a system trying to detect swimming shipwreck survivors). Most of what you see is background.

backscatter   That part of light projected into a volume that is not absorbed or transmitted but scattered back to its starting point. It may be used to describe the part of the light scattered from a surface exposed to laser light, with the proviso that the backscattered light is that which is scattered toward the illuminating source.

baffle   A structural shield that prevents unwanted light from impinging on the focal plane. Cold shields and sunshades are forms of baffles.

bandgap   The energy bandgap in a semiconductor material is the amount of energy needed to sufficiently interact with the lattice to generate free carriers, which are the source of the detected electrical signal. For photovoltaic and photoconductive devices, the bandgap defines the longest wavelength to which the detector is sensitive.

bandpass   The range of wavelengths over which a system functions. The term is also used in an electronic sense to define the range of electrical frequencies that are accommodated by an electrical system.

Bessel   A prominent mathematician whose work has been recognized by assigning his name to a function frequently used in defining the electromagnetic properties of circular apertures, such as in diffraction theory.

binary optics   Optics that utilize diffraction to alter a light ray. Binary optics are made from a photolithography (mask and etch) process. The photolithographic process results in the optical curve being approximated in a series of steps. This staircase approximation to a curve has the number of steps being a power of 2 (e.g., 2 steps, 4 steps, 16 steps, etc.) hence the *binary*.

blackbody   A blackbody has the properties of complete absorptance and emits exactly as described by Planck's Law.

BLIP   *Background limited in performance.* A BLIP system is one in which the noises are dominated by the shot noise of the background. In such cases, reducing the noise of the detector will not provide increased system sensitivity.

blow-down   A cooling system, usually employing a highly pressurized gas expanded through an orifice to generate cooing. *See also* JT (Joule-Thomson).

bolometer	Any of a class of detectors that rely on temperature change in the detector to indicate the presence of radiation, usually infrared.
BQ	*Beam quality.* A measure of the ability of a laser beam expander to form a beam of ideal (that is, diffraction-limited) quality. Exact definitions depend upon application.
Cassegrain	Any of a class of two mirror telescopes, named after Guillaume Cassegrain, in which the primary mirror is concave of a parabolic shape, and the secondary mirror is convex and hyperbolic. These systems exhibit no spherical aberrations.
CCD	*Charge-coupled device.* A particular implementation of detector and/or readout technology composed of an array of capacitors. Charge is accumulated on a capacitor and shift-registered along rows and columns to provide a readout. These are frequently used in commercial video cameras, visible arrays, and as a readout multiplexer for infrared focal planes.
CdS	*Cadmium sulfide.* A detector material still in use but whose application in advanced EO systems has been nearly eliminated by advancements in other detector materials.
centroid	The resultant "center of mass" or other characterization of a distribution of light falling on an array of detectors. In the context of this book, it usually refers to light on a focal plane, often used to compute the location of a target to higher resolution than can be achieved by merely using the location of the distribution of light.
charge skimming	A technique to remove a portion of the noise-generated photoelectrons from a multiplexer. It allows longer integration times.
Chretien	Used generally with the name *Ritchey* as in *Ritchey-Chretien.* Often used to describe an optical implementation of a Cassegrain two-mirror telescope that has the property of having no spherical aberration or coma. This is achieved by using hyperbolic surfaces for both the primary and secondary mirrors
CIE	Commission Internationale de l'Eclairage, or the International Commission on Illumination, a standards-generating group concerned with color and illumination.
clutter	The non-Gaussian spatially and intensity variations in the background. This stuff is hard to filter out and usually looks like what you are trying to detect.
CMOS	*Complementary metal oxide semiconductor,* a common architecture for integrated circuits.
C_n^2	The index of refraction structure constant, pronounced "see en squared." This parameter is widely used in the description of the impact of atmospheric turbulence on the propagation of light in the atmosphere.

COBE	*Cosmic background experiment,* a satellite built to explore the background radiation of the universe.
coefficient of performance	A unitless figure of merit used by refrigeration engineers. Generally, it is the delivered cooling capacity in watts divided by the electrical input in watts.
cold body	A target that is not thrusting or expelling any hot gas. When launching, the Space Shuttle is not a cold body, but when orbiting it is.
cold shield	A cold baffle (usually near an internal reimaging plane and the FPA). Its function is to limit the acceptance angle of the FPA to that of the final F-cone, thereby reducing the contribution of unwanted warm radiation and reflections, from the surrounding housing, to the noise.
common module	A series of electro-optical standardized subsystems (including detectors and dewars) sponsored by the U.S. Army. These form the basis of the vast majority of IR systems fielded by the U.S.A. prior to 2000.
confocal	The property of two or more optical components having the same focal point. Commonly used in describing certain types of laser resonators.
CRT	*Cathode ray tube.*
cryocooler	A refrigerator for providing cooling to cryogenic temperatures ($\approx$ less than 200 K). These include a host of mechanical cycles, gas expansion systems, Peltier effect cooling systems, magnetic cooling systems, sorption systems, and others.
cryostat	The cold finger expansion assembly of a Joule-Thomson cooler. At the tip of the cryostat, the gas is expanded, and cooling power is provided. Generally, a cryostat includes the expansion valve, a counter current heat exchanger, some structure, and mounting surfaces.
D*	This is the abbreviation for the specific detectivity, pronounced "dee-star." This is a figure of merit for detectors defined as

$$\frac{\sqrt{(A_d \Delta f)}}{NEP}$$

where A_d is the area of the detector, Δf is the noise bandwidth, and *NEP* is the noise equivalent power. D* normalizes most detector noises and can be a powerful figure of merit. It generally represents the signal-to-noise ratio produced by 1 W of input power to a detector. Unfortunately, D* depends on the measurement procedures, and frequently important noise mechanisms are omitted (such as $1/f$ noise).

Dall

Usually used with Kirkham, such as in describing the Dall-Kirkham implementation of the two-mirror Cassegrain optical system.

dewar

A double-walled vacuum vessel providing thermal isolation. In the context of this book, it is the evacuated mechanical container that contains the cold train (usually cold finger, detector, cold shield, and cold filter). The term is also used to indicate the storage device (dewar) for liquid cryogens.

die

An integrated circuit as discriminated on the wafer; the chip as delineated on the wafer.

distributed aperture

A system that has more than one aperture feeding light or information to a common focal plane or processor. An example would be five television cameras scattered around a security installation, each feeding a centralized processor that automatically identifies targets.

downwelling

In the context of this book, a directional definition commonly used in describing the flow of electromagnetic radiation in the ocean and atmosphere, with "down" referring to a direction toward the center of the earth. Contrasts with upwelling.

EFL

Effective focal length. The resultant focal length of a system of lenses or mirrors that are working in conjunction.

emissivity

The unitless ratio of a surface's exitance to that of a perfect blackbody. A true blackbody has an emissivity of one, and gray bodies have emissivities of less than one. A shiny object (e.g., bare metal) has low emissivity.

emittance

Synonymous with exitance, which is flux (energy per unit area) leaving a surface.

ensquared

Usually used to describe the amount of energy focused into an area (such as a detector).

EO

Electro-optics. The field of study that integrates electronics and optics to create systems.

erf

The *error function,* defined in any book on mathematical physics or engineering physics.

ERIM

The Environmental Research Institute of Michigan. A leading center for remote sensing and other areas of electro-optical science.

etalon

A type of interferometer in which high spectral resolution can be obtained. The Fabry-Perot interferometer is such a device.

Etendue

The product of an optical system's solid angle field of view and its effective collective area.

euphotic

Usually used in conjunction with the word *zone.* Refers to that part of the ocean where light plays a role in the biology. This is commonly taken to be the range of depths where the intensity of sunlight is at least 1 percent of the value just under the surface.

exitance	Flux density emitted.
f#	Pronounced "f number." It is commonly defined as the ratio of the effective focal length to the effective aperture.
Fabry	An optical scientist of about 100 years ago whose name is most often seen with that of Perot, as in the Fabry-Perot interferometer.
false alarm rate	The frequency at which a system generates false alarms. Related to the probability of detection and the amount of noise and clutter.
figure	The general overall shape of an optic (e.g., a parabola, sphere, and so forth).
Fizeau	A prominent early optical scientist whose name is associated with certain types of interferometers.
FLIR	*Forward-looking infrared.* A class of sensors often used in defense and surveillance applications
focal length	The distance from the principal plane to the focal plane.
focus	The act of forming the best possible image.
FOR	*Field of regard.* The field over which a sensor can be operated. Differs from *field of view* in that the latter is the field that can be observed without moving the instantaneous line of sight. The FOR commonly describes the field that can be covered using gimbals or steering mirrors in the system.
Foucault	An important scientist who contributed to the study of both mechanics and optics; in the latter case, especially on the problem of determining the speed of light. The knife edge test is named after him.
Fourier	A prominent scientist and mathematician who gives his name to any number of important fields of current study. In electro-optics, his name appears associated with the field of Fourier optics, the Fourier transform, and so on.
FOV	*Field of view.*
FPA	*Focal plane array* (or assembly). An integrated electronic package that includes detectors and multiplexing readout electronics.
frame differencing	An image processing technique that compares one frame (usually by subtraction) of a given scene from one some time later. This is a very effective way of reducing clutter, and detecting moving targets, changes to the scene, and new targets.
Fried	David Fried. A prominent participant in the development of the modern theory of radiation propagation in turbulent me-

dia. Gives his name to the "Fried parameter," a measure of the optical properties of the atmosphere.

Fried parameter

A measure of the distorting effect of the atmosphere as the result of light propagating over a path in which turbulence is present. Named for David Fried, a prominent optical scientist whose work defined the field.

FWHM

Full width, half maximum. A common measure of the width of a distribution of some function. For example, in laser physics, this term is often used to define the lateral dimension of a beam by determining the diameter that best meets the criterion of being at the half-power point of the beam.

GaAs

Gallium arsenide. A semiconductor material used in electronics, optics and detectors.

GHz

Gigahertz = 10^9 (1 billion) hertz.

graybody

A surface or object that has an radiant exitance with the same distribution as a blackbody, but with lower intensity.

Hartmann

An optical researcher from the turn of the century who contributed the concept of a multi-aperture lens system that is used in optical testing and in wavefront sensing. *See also* Shack-Hartmann (or Hartmann-Shack) sensor.

HeNe

Short for a *helium neon* laser. HeNe lasers are common, low in cost, and commercially available. They typically have a frequency of 0.6328 µm.

HgCdTe

Mercury-cadmium-telluride is an alloy that provides photoconductive and photovoltaic conversion of light from about 1 to 22 µm.

Hubble

The famous American astronomer after whom the Hubble space telescope is named.

HUD

Head-up display.

IFOV

Instantaneous field of view. The field of view of a single pixel in a multidetector array.

illuminance

Flux that is incident on a surface.

in-band

Having the property of being detectable by a particular electro-optic system. That is, the radiation is said to be within the wavelength limits of the system.

index of refraction

A property of all transparent or semi-transparent materials, defined by the speed of light in the material and, in turn, defining the effect of the material on the deviation of light entering from another medium of different index.

infrared

The portion of the electromagnetic radiation spectrum that extends in wavelength from beyond the visible ($\approx$0.8 µm) to the submillimeter regime ($\approx$100 µm).

InGaAs

Indium gallium arsenide.

InSb	*Indium antimonide.* A common mid-infrared detector material. Pronounced "ins bee."
insolation	The amount of radiation imposed on a surface by the sun.
integral Stirling	A Stirling cooler of an architecture that has the expander directly connected to the compressor.
Interferometry	The field of study for the development and application of systems that work by combining separate beams of light.
IRAS	The *Infrared Astronomical Satellite.*
IRIA	*Infrared Information Analysis Center,* a U.S. government sponsored professional society that supports symposium and information of all subjects pertaining to this book. The address for their home page on the World Wide Web is: http://www.erim.org/IRIA/IRIA.html
IRIS	*Infrared Information Symposium.* A symposium sponsored by IRIA (see above).
isoplanatic angle	The range of angles over which the wavefront of a system is essentially constant.
JT	Abbreviation for Joule-Thomson. A type of cooling system in which high-pressure gas (usually argon or nitrogen of several thousand lb/in^2) is evacuated through an expansion valve to produce a cooling effect.
K	The abbreviation for Kelvins, the internationally accepted unit of temperature.
kapton	A plastic material used in specialized tape, some circuit boards, and some multilayer insulation materials.
Keck	A foundation that gives the name to the 10-m multi-mirror telescope on Mauna Kea.
Kelvins	The internationally accepted metric unit of temperature and the most common measure of temperature in the electro-optics industry. Zero Kelvin is absolute zero, and each Kelvin increment is the same as a degree centigrade. Note that the internationally accepted use of the term eliminates the word degrees, as in "the instrument operates at 10 Kelvin (or Kelvins)."
kilopascals	Measurement unit for pressure. In the most recent definition of pressure, a standard atmosphere is defined as 110,325 Pa or 110.325 kPa.
Kirkham	Usually used with Dall, as in the Dall-Kirkham implementation of the Cassegrain two-mirror telescope.
Kolmogorov	A prominent Soviet mathematician whose work on the statistical properties of the atmosphere forms the foundation of most modern theories of the impact of turbulence on light propagation.

LADAR	*Laser radar.* A system for determining the range, and sometimes shape, of distant targets by illuminating them with laser and light and detecting the reflected light with a co-located detection system. Used interchangeably with LIDAR by most researchers, with the latter meaning "light detection and ranging."
lambert	A unit of light fluence in the photopic/scotopic system of measurement. A lambert is defined as 3183.1 candles/m^2.
least significant bit	The smallest amount of information that is represented when an analog system is digitized. At best, it is about 1/2 of a bit from the A/D. This can be the dominant noise source for quiet, large dynamic range sensors.
LED	*Light emitting diode,* an important technology for producing the light used in communications and displays.
lenslets	Usually used to define the properties of the lens systems employed in Shack-Hartmann wavefront sensors.
lightweighting	The act of removing unnecessary mass from optical systems, usually with the intention of making the instrument compatible with an aircraft or spacecraft platform or to make the system more adaptable to a gimbal mount.
line of sight	The projection of the optic axis into object space. This represents the center of the field of view and where the sensor is pointed.
LiTaO$_3$	A compound (of *lithium, tantalum and oxygen*) that is useful for detection of infrared radiation employing the pyroelectric effect.
LOS	*Line of sight* (see above).
LOWTRAN	LOWTRAN is an acronym for *low-resolution transmission.* It is a modeling tool used to determine the transmission and scattering properties of the atmosphere for a variety of conditions, including season, weather conditions, place on the Earth, and so on. It is generally limited to resolution of 20 cm^{-1}, with extrapolation to 5 cm^{-1}. MODTRAN and HITRAN are similar programs with better resolution. PCTRAN is a version for the personal computer.
LSB	*Least significant bit* (see above).
lux	A measure of radiation in the photopic/scotopic radiation measurement system. Abbreviated lx.
LWIR	*Long-wave infrared.* The portion of the electromagnetic spectrum that is longer in wavelength than mid-infrared and shorter in wavelength than very long infrared. Typically defined by the available detector technology to be from about 6 μm (where Pt:Si and InSb don't work) to ≈ 26 μm (where, typically, most doped silicon fails to work). Often, commercial

and tactical military engineers will loosely use LWIR to mean the 8- to 12-µm atmospheric window.

magneto-strictive	A phenomenon where a material changes dimensions as a function of magnetic fields (similar to electrostrictive and piezoelectric in applications).
magnitude	A measurement of stellar brightness. Abbreviated Mv for visual magnitude.
manufac-turability	The measure of the ease with which something can be manufactured. It is usually included in the description of the risk associated with a particular EO system or technology since it implies a cost in either time or money.
meridianal	A plane through an optical system that includes the optic axis.
microlens	A tiny lens, usually mounted directly to a focal plane.
micron	A millionth of a meter (i.e., a micrometer).
microradian	A millionth of a radian. Often also appears as µrad.
millibars	A measure of pressure, now replaced by Pascals
milliradian	A thousandth of a radian. Often also appears as mrad.
milliseconds	A thousandth of a second.
milliwatt	A thousandth of a watt. Often appears as mW.
MODTRAN	An optical modeling tool that is used to estimate the transmission and scattering properties of the atmosphere. It is used in cases where the spectral resolution of LOWTRAN is insufficient. Generally limited to resolution of 1 cm^{-1}.
MRT	*Minimum resolvable temperature,* a common figure of merit describing the performance of a thermal imager. Usually, this implies that the effect of the display and human eye are considered.
MTF	*Modulation transfer function,* a way to describe image quality.
MTTF	*Mean time to failure.* A statistical measure indicating the reliability of a system at which time half of the units have failed.
multimode	Generally, this refers to fiber optics or lasers and indicates that they are capable of transmitting (in the case of fibers) or generating (for lasers) radiation of more than one electromagnetic mode.
multiplexer	An electronic system able to provide inputs from several sources to one operational module, in turn. Often used in describing the design of electronic focal plane readout systems.
mux	Multiplexer.
mW	Milliwatt.
MWIR	*Mid-wave infrared,* usually includes radiation from about 3 to 6 µm.

mylar	A plastic film material often used in the fabrication of multilayer insulation materials for cryogenics and spacecraft.
NA	*Numerical aperture.*
nanoampere (nanoamp)	One billionth of an ampere.
NEDT	*Noise equivalent delta temperature,* a figure of merit suitable for infrared thermal imagers. Also abbreviated NEΔT.
NEI	*Noise equivalent irradiance.* A measure of the performance of a system in which all of the noise terms are aggregated into one measure that equates with the irradiance to a noise-free system. Defined in units of watts/cm^2.
NEP	*Noise equivalent power.* A measure of the performance of a detector or system in which all of the noise terms are aggregated into one measure that equates with the power needed to equal them. NEP is defined with units of watts.
newton	A metric unit of force.
nm	*Nanometer.* A commonly used measure of wavelength defined as one billionth of a meter.
NOAA	National Oceanic and Atmospheric Administration (U.S.A.).
nonisotropic	Not having the properties of isotropy; that is, not being the same in all directions of propagation.
nonthrusting	Describing the properties of a target that is not equipped with jet or rocket engine. Often used in characterizing the signature properties of missiles whose engines have burned out or the properties of the product of a rocket, such as a re-entry vehicle, that is not equipped with engines.
ns	*Nanosecond.* One billionth of a second.
Nyquist frequency	In sampling theory, this is the highest frequency that can be faithfully reproduced and is equal to two times the system resolution. Also called the *Nyquist criterion.*
obscuration	An opaque blockage to light in the optics of a system.
OPD	*Optical path difference.* Used in describing the performance limitations of optical systems in which the phase front of the light propagating in the system is aberrated by the optics or the turbulent environment outside of the optics.
optical depth	The integration of the absorption coefficient along the path of light in an absorbing medium.
optical element	A lens, mirror, flat, or piece of optics in an optical assembly.
optical gain	The effective radiometric aperture divided by the blur spot size.
optomechanics	The field of engineering that addresses the integration of optics and mechanical structures and mechanisms.

overscan	To scan more than is strictly needed, usually to provide a margin of error for pointing or to viewing a radiometric reference.
PbS	*Lead sulfide.* An important infrared detector material providing short wave response. One of the first IR detector materials made, it provides good sensitivity with thermoelectric cooling.
PbSe	*Lead selenide.* An important infrared detector material providing short and mid wave photoconductive sensing. Like PbS, it was one of the first IR detector materials made, it provides good sensitivity with thermoelectric cooling.
PC	*Photoconductive.*
PDR	*Preliminary design review.* A meeting at which early plans for system development are presented and reviewed.
Peltier	The discoverer of the basic physics that has led to the development of the thermoelectric cooler. A type of cryogenic cooler.
P_{fa}	*Probability of false alarm.* A term often found in discussions of detection theory and the design and performance assessment of such systems.
phenomenology	In the context of this book, the characteristics of chemical reactions that yield electro-optic signatures from plumes, targets, and background.
photoemission	A process of the emission of an electron from a surface when a photon encounters it. This is the basis of operation of the photomultiplier tube.
photoconductive	A type of detector that changes conductance (resistance) on exposure to light. Contrasts with the photovoltaic detector types.
photodetectors	A broad description of components able to sense the presence of light in a quantitative way. Generally, this category includes photomultipliers, photodiodes, CCDs, and so forth.
photolithographic	A process using masks and etching to create very minute (less than a micron is common) features. It is widely used by the electronics industry to create integrated circuits.
photomultiplier	A detection system making use of the photo-electric effect and the amplification of the resulting electron emission from the photocathode to provide extremely sensitive non-imaging sensors.
photon	A massless bundle of energy traveling at the speed of light, associated with light and light phenomena.
photonics	The general field of creation, detection, and manipulation of photons of light. Includes virtually all of the modern electro-optic sciences
photopic	Characterizing the response of the cones in the human retina. *See also* scotopic

photovoltaic	A type of detector that produces an output voltage difference between two electrodes in response to impinging light.
phytoplankton	Small plants, usually unicellular, that participate in the absorption, scattering, and fluorescence properties of the ocean.
pico	A fraction of 10^{-12}, as in "a picowatt equals one millionth of one millionth of a watt."
pixel	Formally, a picture element, and should strictly be only applied to displays. However, it is commonly used to refer to an individual detector on a focal plane array.
Planck function	Refers to Max Planck's blackbody radiation law.
plume	The exhaust emission from a chemical process. In this book, it refers to the exhaust cloud from a car's tailpipe, the exhaust from a jet engine, and the flames and exhaust from a rocket engine.
power spectral density	A mathematical representation of the variations in a measurement plotted in a frequency-like domain. In the context of this book, it is used to describe scene clutter or vibration.
probability of detection	A statistical measurement of the likelihood that an object can be detected in the presence of noise.
probability of false alarm	A statistical measurement of the likelihood that a noise source will be identified as a target.
producibility	The measure of the ability and effort required to produce a particular item (system or component) at a given production rate.
projected area	The effective two-dimensional silhouette of a three-dimensional object. A sphere has a projected area of a circle of the same radius.
PSD	*Power spectral density.*
Pt:Si	*Platinum silicide.* A commonly used Schottky barrier device used as an infrared detection material. Typically characterized by low noise, relatively low quantum efficiency, but high uniformity and ease of producibility.
pulse tube	A variant of a Stirling cooler that replaces the piston in the expander with a "slug" of gas.
PV	When used with detectors, *photovoltaic* (see definition above). When used with optics, *peak to valley.*
pyroelectric	A type of detection material that is slow to respond but has a very wide spectral sensitivity range.
quantum efficiency	When used in detectors, this is a figure of merit describing the conversion efficiency from a photon to a usable free carrier (electron or hole). 100 percent quantum efficient implies that all the photons incident generate carriers that can be read out.

QWIP	*Quantum well infrared photodetector* (or *photoconductor*).
radiometrics	The science of measuring the flow of electromagnetic energy in quantitative terms.
radiometry	The study of the emission, transmission, and absorptance of radiant energy.
rangefinder	Active or passive optical systems capable of measuring the distance between points. A laser radar (LADAR) performs this function by illuminating the target and measuring the time delay until the reflected light appears at a sensor co-located with the transmitter. Many other systems can be employed to make range finding measurements.
raster	A pattern generated by scanning one line followed by another.
Rayleigh scattering	A scattering process named after the theory developed by Lord Rayleigh (John William Strutt) to describe the scattering properties of the atmosphere. Generally, Rayleigh scattering effect decreases as wavelength increases (which is why the sky is blue and not red).
refractivity	The same as index of refraction.
reimaging	An optical system that has more than one optical focus region (a focal plane in the optical context, but not in the detector hardware context) and forms images are more than one location. Reimaging systems are common in cryogenic systems and are used with cold shields to limit the impingement of stray thermal emission.
RF	*Radio frequency.*
Ritchey	Used generally with the name Ritchey as in Ritchey-Chretien. This is a special implementation of a Cassegrain two-mirror telescope that has the property of having no spherical aberration or coma. This is achieved by using hyperbolic surfaces for both the primary and secondary mirrors.
r_0	A common abbreviation for Fried's parameter.
R_0A	The resistance area product. The "o" refers to the measurement taken at zero bias. This is a handy semiconductor figure of merit for evaluating detector material quality. This measurement is easily taken. For detector applications, this allows calculation of the Johnson noise and relates to sensitivity (the higher the R_0A, the better it is). It is very temperature dependent, (e.g., can change several orders of magnitude in InSb for a 10 or 20 Kelvin change). It is pronounced "are not a."
Rytov	A theoretical physicist prominent in the development of the theory of radiation propagation in turbulent media. His name is connected with a limit to the amount of turbulence that can be present before the "weak turbulence" theory does not apply and more complex modeling methods must be employed.

saccades — Small motions of the human eye.

Schmidt — A telescope design that includes a correcting refracting plate to provide high quality imaging over a large field.

Schottky barrier — A class of semiconductor technology and process. In this book, it refers to a class of detectors that work by internal semiconductor photoemission. Examples are Pd:Si, Pt:Si, and Ir:Si.

scotopic — Characterizing the properties of the response of the rods in the human retina. See also *photopic*.

SCR — *Signal to clutter ratio* (see below).

seeing — A measurement of the viewing "goodness" at a given place and time. This is frequently used by astronomers and relates to the turbulence present.

SEM-E — *Standard electronics module.* A standardized circuit board size and I/O architecture, popular with American military systems.

semiconductor — A material with a concentration of free carriers that allows it to be a conductor or resistor and allows this property to be controlled by an applied voltage.

Si — *Silicon.* The most commonly used material in the development of semiconductors. Also used as a material for visible and near infrared detectors and, when doped, short-, mid-, and long-wavelength infrared focal planes. Truly nature's gift to us all.

SiC — *Silicon carbide.* A ceramic material of considerable interest in the development of modern optics and optical structures, particularly because of its important mechanical properties and light weight.

signal-to-clutter ratio — The ratio of the target signal (say in photogenerated electrons) to the signal (again in electrons) caused by sensing the clutter.

slewed — Pointed, as in the slewing of a sensor to increase its field of regard. Usually requires that the entire instrument be repointed, such as by using a gimbal.

SNR — *Signal-to-noise ratio.*

SPIE — *Society of Photo-Optical Instrumentation Engineers.* An international professional society which deals of all subjects pertaining to this book.

SPIRITS — *Spectral infrared targets and scenes.* This is a software modeling tool to estimate spectral radiance from targets and backgrounds.

split Stirling — A Stirling cooler of an architecture that has the expander separate from the compressor and is powered by pressure differentials in a gas line.

SPOT — *Satellite Probatorie d'Obervation de la Terre* (A French remote sensing system).

SRR	*System requirements review.* An early meeting in the development of a system during which the requirements for the system that have been derived from performance requirements can be reviewed and improved.
steradian	The unit of measure of solid angle. Computed as the ratio of the area of the cap of a sphere to the square of the radius to the point at which the solid angle is measured.
stilb	A measure of light in the photopic/scotopic system that is equal to 10,000 candles/m^2.
Strehl	An optics researcher who lends his name to the oft-quoted measure of the performance of an optical system in which the intensity that is projected or collected is compared with that of a perfect system.
subaperture	Often used in discussing the properties of the Hartmann-Shack wavefront sensor, which is made up of many independent but adjoining optical systems, each equipped with its own lens.
subpixel	Usually used in describing the performance of multidetector systems in locating targets to an accuracy better than can be achieved by the field of view of a single detector pixel alone, usually by calculating the "centroid" of the light falling on the detectors or observing the rise time of the signal as the blur spot crosses pixels.
subtense	The angular extent of a system, such as the field of view of a sensor.
superresolution	The process of providing, by analysis, information on the location or shape of a target that exceeds that which can be obtained by the optical system alone. Usually implies a positional accuracy better than the diffraction limit. *See also* subpixel.
SWIR	*Short-wave infrared.* The part of the electromagnetic spectrum with wavelengths longer than the near infrared ($\approx$1.1 μm) and shorter than mid infrared ($\approx$3 μm).
tanh	Hyperbolic tangent function.
TDI	Time delay and integration (see below).
TE	*Thermoelectric,* as in a thermoelectric cooler.
TEC	*Thermoelectric cooler.*
Temporal processing	Signal or image processing that operates on features in the time domain. Frame-to-frame differencing is an example.
thermocline	In oceanography, the layer of water in which the change in temperature with depth is greater than the change in the overlying or underlying layers. This type of structure can have a profound effect on the optics of the ocean.
time delay and integration	A signal-enhancing process whereby a scan is synchronized to the integration of several detectors in the scan direction. As the blur spot moves from one detector (in the scan direction) to

another, the signal is added (usually in the analog domain). The signal to noise equal to the square root of the number of TDI detectors in the scan direction.

TIS   *Total integrated scatter.*

Torr   A measure of pressure equal to 133.32 Pascals, the latter unit being a more common modern unit.

tracker   An electro-optic system designed to locate, follow, and report on the position of a target. The information from the tracker is often used to control some other system, such as an astronomical telescope, weapon system, or the like.

transmission   The measure of the amount of radiation that passes through a substance.

transmissive   Having the property of being partially transparent. Must be specified as to wavelength.

transmittance   The ratio of the amount of radiation of a particular wavelength that passes through a path in a material to the amount incident on the material.

TTL   *Transistor-to-transistor logic* (usually operates at 3 to 5 V).

turbulence   In atmospheric optics, the presence of cells of index of refraction that vary from place to place and time to time, causing refraction of light rays, ultimately limiting the ability of images to be formed.

uncooled   Systems or components in which operation is adequate at the ambient temperature of the surrounding system. Or, systems or components in which no special effort is made to include cooling systems such as cryogenic fluids or gases, refrigerators, or other mechanisms.

unitless   Having properties such that the parameter is not associated with any particular measure of distance, volume, time, etc. An example is solid angle which, although it is often defined in terms of steradians, is the ratio of two lengths squared.

unobscured   Often used to describe optical system designs in which there are no structures or other nontransmissive parts between the system and the object being imaged. Off-axis mirror telescopes and most refracting optical systems satisfy this definition, while the common on-axis two mirror telescope does not.

upwelling   A directional definition commonly used in describing the flow of electro-magnetic radiation in the ocean and atmosphere, with "up" referring to away from the center of the earth. Contrasts with downwelling.

UV   *Ultraviolet.* This is part of the electromagnet spectrum with wavelengths longer than X-rays (≈ 0.1 μm) but shorter than (≈ 0.4 μm) visible light.

vidicon	An image collection technology that uses a scanning electron beam working in conjunction with a photoconductive detector to form an image.
visibility	The quantitative measure of the ability of human observers to see standard objects at distance through atmospheres containing rain, snow, or other obscurants. Normal visibility is 23 km.
visible spectrum	The part of the electromagnetic spectrum that is approximately visible to the human eye. Generally, this is from about 0.35 to 0.76 µm.
VME	*Versamodule Europe.* A standardized circuit board size and I/O architecture, popular for commercial systems.
wafer	A slice of a boule, usually used to delineate a material disk suitable for semiconductor processing.
wave	A periodic undulation in a field.
wave number	A measure of the frequency of electromagnetic radiation. Computed by the formula $1/\lambda$.
wavefront	Characterization of a beam of light, usually for the purpose of characterizing the degradations induced by passage through a non-perfect optical system or transmission medium.
waveguide	A pipe or other conveyance able to efficiently aid the transmission of light or infrared radiation from one point to another. Fiber optic cables meet this definition.
wavelength	The distance from one peak in a field of waves to the next.
WFE	*Wavefront error,* such as describes the degraded properties of light propagating through other than perfect optics.
WFOV	*Wide field of view.*
white noise	Noise that is not frequency dependent (has a flat PSD).
ZnS	*Zinc selenide.* A common IR refractive material
µrad	*Microradian.* A millionth of a radian.
µW	*Microwatt.* A millionth of a watt.

Appendix A
Useful Values and Conversions

The following idiosyncratic collection of values represent frequently needed information and conversion factors for the EO practitioner. This eclectic collection consist of constants, conversions, and definitions categorized by the authors' whims. Although supplied, the authors are not encouraging the use of discouraged and archaic units. Such units are included here only as a help with their translation to more accepted SI units.

Angular Measurement

Arcseconds	4.848 microradians
	2.77×10^{-4} degrees
Degree	0.01745 radians
	17.5 milliradians
	60 minutes
	3600 seconds
	17,452 microradians
Minute	0.0167 degrees
	2.909×10^{-4} radians
Radian	0.159 of the circumference
	57.296 degrees
	3438 minutes
	2.06×10^5 seconds
Steradian	0.08 of total solid angle (4π)
Radians per second	0.159 revolutions/second
	9.55 revolutions/minute
	57.3 degrees/second

RPM	6 degrees/second
	0.0167 radians/second
	0.105 radians/second
RPS	21,600 degrees/minute

Area Measurements

Square centimeter	1.076×10^{-3} feet^2
	0.155 inches^2
	1×10^{-4} meters^2
Square mil	645 microns^2
Square inch	6.452 centimeters^2
Square foot	939 centimeters^2
	0.092 meters^2

Astronomy

Astronomical unit (mean Earth–Sun distance)	1.496×10^8 kilometers
	93 million miles
Light year	9.46×10^{15} meters
	9.46×10^{12} kilometers
	5.88×10^{12} miles
Parsec	3.26 light years
	3.09×10^{13} kilometers
Effective solar temperature	5900 K
Solar constant	1350–1390 watts/meter^2 (mean above atmosphere)
Irradiance of a zeroth magnitude star	3.1×10^{-13} watts/centimeter^2
Astronomical visual absorption	≈ 0.2 mag/airmass

Atmospherics

Absorption of CH_4	Bands centered at: 3.31 microns, 6.5 microns, and 7.6 microns
Absorption of CO_2	1.35–1.5 microns
	1.8–2.0 microns
	≈ 4.2–4.6 microns
	≈ 14–16 microns
Absorption of H_2O	1.35–1.5 microns
	1.8–2.0 microns
	≈ 2.7–3.0 microns

	5.6–7.1 microns (with the main absorption in ≈6.1–6.5 μm, and some minor narrow bands centered at 0.94 μm, 1.1 μm, 1.38 μm, and 3.2 μm)
Absorption of NO_2	3.9 microns
	4.5 microns
	7.7 microns
	17.1 microns
	and various bands in the UV
Absorption of ozone	≈0.15–0.3 (peak at ≈0.26 microns)
Atmospheric pressure	101,325 newtons/meter^2
	101 kilopascals
	760 mm of Hg at sea level
Density of air @ STP	1.29×10^3 g/centimeter^3
	1.29 kilograms/meter^3
Troposphere altitude (nominal)	0 to approx. 11 kilometers (depends on season and latitude)
Stratosphere (nominal)	11–24 kilometers (some define the stratosphere to include the mesosphere)
Mesosphere (nominal)	24–80 kilometers
Thermosphere (nominal)	80 to approx. 7000 kilometers
Pressure of standard atmosphere	1.01×10^5 newtons/meter^2
	14.7 pounds/inch^2

Cryogens

Boiling point of air	85–88 K
Boiling point of argon	–185.7° C
	87.3 K
Boiling point of Freon–14 CF4	145 K
Boiling point of helium	4.2 K (at standard atmospheric pressure)
	–272.2° C
	≈0.7 K (when area above fluid surface approximates a vacuum)
Boiling point of hydrogen	20.3 K
	–253° C
Boiling point of neon	27.1 K
Boiling point of nitrogen	77.2 K
	–196° C
Boiling point of oxygen	90.2 K
	–218.4° C
Freezing point of water	273.16 K

Room temperature	$\approx 70°$ F
	$\approx 25°$ C
	≈ 300 K
Sublimating dry ice	195 K
Gas constant	8.32 joules/mole K
	1.98 calories/mole K
Avogadro's number	6.02×10^{23} molecules/mole
Heat of fusion of water (0° C)	79.7 calories/gram

Density Measurements

Density of water (4° C, 760 mm Hg)	1 g/centimeter^3
	1000 kilograms/meter^3
	62.43 pounds/foot^3
	0.036 pounds/inch^3
Grams/cubic centimeter	0.036 pounds/inch^3
	64.2 pounds/foot^3
Pounds/inch^3	27.7 grams/centimeter^3

Earth

Escape velocity	11.19 kilometers/second
Gravitational acceleration	9.81 meters/second^2
	32.2 feet/second^2
Mass	$\approx 6 \times 10^{27}$ grams
Mean ocean depth	3800 m
Radius	6371 kilometers (mean)
	3960 miles
	6378 kilometers (equatorial)
	6357 kilometers (polar)
Surface area	5.1×10^{18} centimeters^2
Land area	1.5×10^{18} centimeters^2
Ocean area	3.6×10^{18} centimeters^2
Volume	1.083×10^{27} centimeters^3
Velocity of Earth in orbit	≈ 30 kilometers/second

Electromagnetic Spectrum (in microns)

Gamma rays	< 0.001 microns
X-rays	0.001–0.02 microns
Ultraviolet	≈ 0.02 to approx. 0.4 microns

Visible	≈0.4 to approx. 0.75 microns
Near infrared	≈0.75 to approx. 1.2 microns
Short-wave infrared	≈1.2–3 microns
Mid-wave infrared	≈3–6 microns
Long-wave infrared	≈6–14 microns
Far infrared	≈14–100 microns
Submillimeter	≈100–1000 microns
Radio frequency	>1000 microns

Energy

# Photons in a watt	$5.03 \times 10^{18} \times \lambda$ (with λ in microns)
BTU	252 calories
	1055 joules
	0.29 watt-hours
Calories	4.184 joules
Energy of one electron volt	1.602×10^{-19} J
Energy per photon	$1.98 \times 10^{-19}/\lambda$ watts seconds, where λ is in microns
Erg	1×10^{-7} joules
	2.78×10^{-11} watt hours
Joule	9.48×10^{-4} BTU
	0.2388 cal
	1×10^{7} erg
	1×10^{7} dyne centimeter
	1 watt second
	1 volt coulomb
	0.738 foot pound
	2.78×10^{-4} watt hours
	1 newton meter
	3.73×10^{-7} horsepower hours
Megajoule	1×10^{6} joules
	2.4×10^{7} calories
	A Cadillac traveling 55 mph
Kilogram	9×10^{16} joules

Fiber Optics

Decibels referenced to 1 milliwatt (dBm) to watts	1 picowatt = –90 dBm
	1 nanowatt = –60 dBm
	1 microwatt = –30 dBm
	1 milliwatt = –10 dBm
	10 milliwatt = +10 dBm

Germanium photodiode wavelength range	≈800–1700 nanometers
Indium gallium arsenide photodiode wavelength range	≈900–1800 nanometers
Silicon photodiode wavelength range	≈500–1200 nanometers
Typical fiber windows	0.85 microns 1.3 microns 1.55 microns

Laser Lines (Popular)

Alexandrite	0.72–0.8 microns
Argon	0.51 microns
CO	5.0–7.0 microns
CO_2	9.2–11 microns 10.6 microns popular
DF	3.8–4.0 microns
Doubled Nd:Yag	0.53 microns
Dy:CaF	2.35 microns
Er:Yag	1.64 microns
Erbium	1.54 and 1.57 microns (the popular eyesafe laser)
GaAs	0.9 microns
H_2O laser	28 microns, submillimeters
HeNe	0.6328 microns 0.5944–0.6143 microns 1.152 microns 3.391 microns
HF	2.6–3.0 microns
Kr	0.35 microns
Nd: Yag	1.0645 microns
Ne laser	0.3324 microns 0.5401 microns
Nitrogen laser	0.33 microns
Ruby	0.69 microns
Xe laser	0.46–0.63 microns 2.03 microns 3.51 microns 5.56 microns 9.0 microns

Length

Centimeter	0.3937 inch
	10×10^4 micron
	394 mils
Kilometer	3281 feet
	0.54 nautical miles
	0.621 statute miles
	1094 yards
Meter	1×10^{10} angstroms
	3.28 feet
	39.37 inch
	1×10^9 nanometers
	1.094 yards
Mil	0.001 inch
	25.4 microns
	0.0254 millimeters
Inches	2.54 centimeters
Mile (nautical)	1852 meters
	1.15 statute miles
Mile (statute)	5280 feet
	63360 inches
	160,934 centimeters
	1.609 kilometers
	1609 meters

Miscellaneous

Weight of air	1.2 kilograms/meter^3
False alarm probability (in white noise for a point source) at a given SNR with a P_d of 0.99	SNR of 5, Pfa $\approx$ 0.01
	SNR of 6, Pfa $\approx 5 \times 10^{-4}$
	SNR of 7, Pfa $\approx 1 \times 10^{-5}$
	SNR of 9, Pfa $\approx 1 \times 10^{-10}$
Gaussian probability that a value will not exceed	1 sigma: 68.3%
	2 sigma 95.4%
	3 sigma 99.7%
Index of refraction of water	1.344 @ 0° C
	1.342 @ 30° C
	1.337 @ 60° C
Work month (average)	163 hours
Work week (average)	37.5 hours
Work year	$\approx$2000 hours

Peak of the human eye's response	≈0.4–0.65 microns
Water heat of fusion	80 gram calories
Volume of one mole of gas at STP	22.4 liters

Numerical Constants

e	2.718281828459045
π	3.141592653589793

Optics

Amount of energy in circular diffraction pattern	84% in center disk an additional 7.1% in first bright ring an additional 2.8% in second bright ring an additional 1.5% in third bright ring
optical density ($= \log 10(T)$)	0 = 1.0 opacity and 100% transmission 0.5 = 3.2 opacity and 32% transmission 1.0 = 10.0 opacity and 10% transmission 1.5 = 32 opacity and 3.2% transmission 2.0 = 100 opacity and 1% transmission 3.0 = 1000 opacity and 0.1% transmission
Refractive index of Ge	≈4.0 (be careful of the wavelength dependence)
Refractive index of glass	≈1.5–2.0
Refractive index of quartz	≈1.3–1.5
Refractive index of Si	≈3.4
Refractive index of ZnS	≈2.5

Photonic

Candle per square foot	3.38×10^{-3} lamberts
Foot candle	1 lumen/foot^2 10.76 lumen/meter^2 10.76 lux
Lambert	0.318 lamberts/centimeter^2 295.7 candela/foot^2
Lumen per square foot	10.76 lux
Lux	One lumen/meter^2 1×10^{-4} phot

Physical Constants

Atomic mass unit	1.657×10^{-24} grams
Avogadro's number	6.022×10^{-23} molecules/mole
Boltzmann's constant	1.3806×10^{-23} watt seconds/degree
	1.3806×10^{-23} joules/kelvin
Charge of an electron	1.602×10^{-19} coulombs
Gravitational constant	6.67×10^{-11} m^3/kilogram seconds^2
Permeability of free space	12.566×10^{-7} henries/meter
Permittivity of free space	8.854×10^{-12} farads/meter
Planck's constant	6.6252×10^{-34} watt seconds^2 or
	6.6254×10^{-34} joule seconds
Rest mass of a neutron	1.675×10^{-27} kilograms
Rest mass of a proton	1.673×10^{-27} kilograms
Rest mass of an electron	9.109×10^{-31} kilograms
Velocity of light	2.99793×10^{8} meters/second
	2.99793×10^{10} centimeters/second
	2.99793×10^{14} microns/second

Pressure

Dynes per square centimeter	1.02×10^{-3} grams/centimeter^2
	1.45×10^{-5} pounds/inch^2
One atmosphere	1.0133 bars
	1.013×10^{6} dynes/centimeter^2
	760 Torr
	1033 grams/centimeter^3
	760 millimeters Hg
Ounce (mass)	28.35 grams
Pounds per square inch	6895 pascals
	0.068 atmospheres
	51.71 Torr
	51715 mm Hg
Torr	133.32 pascals
	0.00133 bar
	760 mm Hg

Radiometric

Blackbody constant	2897.9 microns K for peak in watts
	3669 microns K for peak in photons

c1 (radiation constant)	3.7413×10^{-16} watt centimeters^2
	4.993×10^{-24} joules/meter
c2 (radiation constant)	1.4388 centimeter degrees
Stefan-Boltzmann constant	5.6686×10^{-12} watt • centimeter^2 • degrees^{-4}
	1.354×10^{-12} calorie • centimeter^2 • degrees^{-4} • seconds^{-1}

Temperature

0° C	273 K
	492° Rankine
	32° F
0 K	−273.16° C
	−459.7° F
A difference of 1° C	1 K
	1.8° Fahrenheit
	1.8° Rankine
	0.8° re' aumur

Time

Hour	3600 seconds
	0.04167 days
	5.95×10^{-3} weeks
Day (mean = 24 hours)	1440 minutes
	86400 seconds
Month (average)	30.44 days
	730.5 hours
	2.63×10^6 seconds
	4.348 weeks
Year	365.256 days (sidereal)
	8766 hours
	525,960 minutes
	3.16 million seconds
	52.18 weeks

Velocity

Mach 1 (as defined as velocity of sound in air at STP)	345 meters/second
	34,500 cm/second
	1132 ft/second
	771 miles/hour
Meter/second	2.24 mph
Miles/hour	88 feet/minute
	1.467 feet/second

1.609 kilometers/hr
0.869 knots
26.822 meters/minute
0.44 meter/second

Velocity of sound in air
usually ≈330–350 m/second
usually ≈1080–1150 ft/second

Velocity of sound in water
1470 meters/second
4823 feet/second

Volume

Cubic centimeter
1×10^{-6} meter^3
1000 millimeters^3 or 0.061 inch^3
2.64×10^{-4} gallons
3.53×10^{-5} feet^3

Cubic meter
61,024 inch^3
1.307 yard^3
1×10^6 centimeter^3

Liter
1000 centimeter^3
0.0353 foot^3
61.02 inch^3
0.001 meter^3
0.264 gallons (U.S.A.)

Weight

Gram (at 1 g)
2.2×10^{-3} pounds
0.001 kilograms

Ounce (mass)
28.35 grams

Pound (mass)
454 grams

Appendix B
World Wide Web Locations for Information on Photonics

The following is a collection of Internet locations where those who are interested in photonics might find some good and timely information. This was assembled from the authors' favorite places at the time of manuscript submission in late 1995. Many more are likely to appear, and many of these are subject to disappearance; nevertheless, this can be an important starting point and reference for many readers.

Institution	URL address	Notes
Allied Signal, Kansas City Plant	http://www.os.kcp.com/ home/catalog/eleopt-sen.html	
American Astronomical Society	http://blackhole.aas.org/ aashomepage.html	
American Institute of Physics	http://www.aip.org	
American Physical Society	http://aps.org	
Arcetri	http:// www.arcetri.astro.it/ home/idefix/mosaic/ ir_base.html	The activities of Astrofisico di ArcetriCSIRO
Coherent, Inc.	http://cid.cohr.com	
CSIRO	http://www.dap.csiro.au/ optek/	Much information on radiometric calibration
Fiber Optic Marketplace	http://www.fiberoptic.com	Home page
General Electric Corporate Research Center	http://www.ge.com/crd	
Hopkins UV Telescope	http://praxis.pha.jhu.edu/ hut.html	

Institution	URL address	Notes
IRIA	http:/www.erim.org/iria/iria.html	Home page
Johns Hopkins University	http://praxis.pha.jhu.edu/	Provides an index of Johns Hopkins University's *Journal of Electronic Defense*
Journal of Electronic Defense	http://www.jede-fense.com/jed.html	Home page
Laser Focus World	http://www.lfw.com/	
MIT Space Resources	http://space.mit.edu	
NASA MSFC, Electro-Optics Branch	http://eo.mfsc. nasa.gov	Home page
NASA, Ames Research Center, IR Technology Service	http://irteck.arc.nasa.gov	Home page
NASA, Jet Propulsion Laboratory	http://www.jpl.nasa.gov	JPL's home page; much photonics-related material, including about SIRTF
NASA, Langley Research Center, Remote Sensing Thrust Office	http://aesd.larc. nasa.gov/c/cf	
NASA, Marshall Spaceflight Center	http://spacelink.msfc.nasa.gov	NASA's Spacelink via the Marshall Center
National Institute of Standards and Technology	http://physics.nist.gov/majresfac/facilities.html	Entry into NIST's network and much information
NOSC	http://www.nosc.mil/planet_earth/optics.html	
Optical Engineering	http://www.yahoo.com/science/engineering/optical_engineering/	Also check the material science area
Optical Society of America	http://www.osa.org	
Society for Imaging Science and Technology	http://www.imaging.org	
SPIE	http://spie.org	Home page for assortment of topics
Stanford	http://www-star.stam-ford.edu/course	Stanford's Space and Telecommunications Center and courses
Tufts University	http://www.eotc.tufts.edu	The electro-optical technology center at Tufts
UC Irvine, Dept. of Cognitive Sciences	http://www.sosci.uci.edu/cogsci/vision	
University of Rochester	http://www.optics.rochester.edu	
USAF, Phillips Laboratory	http://www.plk.af.mil	Home page

Selected Bibliography

Accetta, J. and D. Shumaker, exec eds. 1993. *The Infrared and Electro-Optical Systems Handbook.* Ann Arbor, MI: ERIM,) and Bellingham, WA: SPIE, ISBN 0-8194-1072-1

Allen, C. 1955. *Astrophysical Quantities.* London: Athlone Press.

Apel, J. R. 1987. *Principles of Ocean Optics.* Orlando, FL: Academic Press.

Augustine, N. 1983. *Augustine's Laws.* New York: Viking. ISBN 0-670-80942-X.

Bamberg, P., and S. Sternberg. 1988. *Mathematics for Students of Physics,* vol. 1. Cambridge, MA: Cambridge University Press. ISBN 052125017X.

Barnoski, M., ed. 1981. *Fundamentals of Optical Fiber Communications.* New York: Harcourt Brace Jovanovich. ISBN 0-12-0779151-X.

Begunov, B. et al. 1981. *Optical Instrumentation.* Moscow: Mir. ISBN 5-03-000008-9.

Born, M. and E. Wolf. 1980. *Principles of Optics.* Oxford, U.K.: Pergamon.

Bramson, M. 1968. *Infrared Radiation.* New York: Plenum Press.

Bright, J., ed. 1968. *Technological Forecasting For Industry and Government.* Englewood Cliffs, NJ: Prentice-Hall.

Burgelman, R., and M. Maidique. 1988. *Strategic Management of Technology and Innovation.* Homewood, IL: Irwin. ISBN 0-256-03481-8.

Burle engineering staff. 1974. *Electro-Optics Handbook.* Lancaster, PA: Burle Industries.

Chaisson, E. 1994. *The Hubble Wars.* New York: HarperCollins. ISBN 0-06-017114-6.

Dereniak, E. and D. Crowe. 1984. *Optical Radiation Detectors.* New York: John Wiley & Sons. ISBN 0-471-89797-3.

Driscoll, W., and W. Vaughan, eds. 1978. *Handbook of Optics.* New York: McGraw-Hill. ISBN 0-07-047710-8.

Ealey, M., ed. 1991. *Active and Adaptive Optical Systems,* vol. 1542 of the SPIE publication series. Bellingham, WA: SPIE. ISBN 0-8194-0670-8

Fisher, D. 1988. *Rules of Thumb for Scientists and Engineers.* Houston: Gulf Publishing. ISBN 0-87201-786-9.

Fowles, G. 1989. *Introduction to Modern Optics.* New York: Dover. ISBN 0-486-65957-7.

Geiger, R., P. Allen, and N. Strader. 1990. *VLSI.* New York: McGraw-Hill. ISBN 0-07-023253-9.

Greenleaf, A. 1950. *Photographic Optics.* New York: Macmillan.

Hecht, E. 1987. *Optics.* Reading, MA: Addison-Wesley. ISBN 0-201-11609-X.

Hecht, E. 1975. *Optics.* Schaum's Outline Series in Science. New York: McGraw-Hill. ISBN 0-07-027730-3.

Hecht, E., and A. Zajac. 1974. *Optics.* Menlo Park, CA: Addison-Wesley. ISBN 0-201-02835-2.

Holst, G. 1995. *Electro-Optical Imaging System Performance.* Winter Park, FL: JCD Publishing. ISBN 0-9640000-1-6.

Horowitz, P., and W. Hill. 1980. *The Art of Electronics.* London: Cambridge University Press. ISBN 0-521-23151-5.

Howard, J., ed. 1986. *Optics Today.* New York: American Institute of Physics. ISBN 0-883-18499-0.

Hudson, R. 1969. *Infrared System Engineering.* New York: John Wiley & Sons. ISBN 4-714-1850-1.

Johnson, B. 1960. *Optics and Optical Instruments.* New York: Dover. ISBN 486-60642-2.

Johnson, R., and Wolfe, E., eds. 1985. *Selected Papers on Infrared Design.* Bellingham, MA: SPIE. ISBN 0-89252-548-7.

Kapany, N. 1967. *Fiber Optics.* New York: Academic Press.

Katzman, M., ed. 1987. *Laser Satellite Communications.* Englewood Cliffs, NJ: Prentice-Hall. ISBN 0-13-523804-8

Kimmitt, M. 1970. *Far Infrared Techniques.* London: Pion. ISBN 8-508-6009-1.

Kingslake, R. 1983. *Optical Systems Design.* Orlando, FL: Academic Press. ISBN 0-124-08660-8.

Klein, M. 1970. *Optics.* New York: John Wiley & Sons. ISBN 4-7149-080-6.

Levi, L. 1968. *Applied Optics.* New York: John Wiley & Sons. ISBN 0-4715-3110-3.

Linde, D., ed. 1990. *Handbook of Chemistry and Physics,* 71st edition. Boca Raton, FL: CRC Press. ISBN 0-849-30471-7.

Lloyd, J. 1975. *Thermal Imaging Systems.* New York: Plenum Press. ISBN 0-306-30848-7.

Longhurst, R. S. 1976. *Geometrical and Physical Optics.* New York: Longman. ISBN 0-582-44099-8.

Lovell, D. 1981. *Optical Anecdotes.* Bellingham, MA: SPIE. ISBN 0-892-52353-0.

Malacara, D. 1978. *Optical Shop Testing.* New York: John Wiley & Sons. ISBN 0-471-01973-9.

Miller, J. 1994. *Principles of Infrared Technology.* New York: Van Nostrand Reinhold. ISBN 4-420-1210-1.

NASA Goddard Space Flight Center. 1972. *Advanced Scanners and Imaging Systems for Earth Observations.* Washington, DC: U.S. Government Printing Office

O'Neill, E. 1992. *Introduction to Statistical Optics.* New York: Dover. ISBN 0-486-67328-6.

Pratt, W. 1969. *Laser Communication Systems.* New York: John Wiley & Sons. ISBN 4-716-9650-1.

Press, W., et al. 1986. *Numerical Recipes.* Cambridge, MA: Cambridge University Press. ISBN 0-521-30811-9.

Rockwell, W. 1971. *The Twelve Hats of a Company President.* Englewood Cliffs, NJ: Prentice-Hall. ISBN 0-13-934166-8.

Rose, A. 1977. *Vision: Human and Electronic.* New York: Plenum Press. ISBN 0-306-30732-4.

Schroeder, D. 1987. *Astronomical Optics.* Orlando, FL: Academic Press. ISBN 0-126-29805-X.

Seigman, A. 1986. *Lasers.* Mill Valley: University Science Books.

Seyrafi, K. 1973. *Electro-Optical Systems Analysis.* Los Angeles: Electro-Optical Research Company.

Shannon, R., and J. Wyant. 1980. *Applied Optics and Optical Engineering.* New York: Academic Press. ISBN 0-12-408608-X.

Smith, W. 1966. *Modern Optical Engineering.* New York: McGraw-Hill. ISBN 0-705-8690-X.

Spiro, I., and M. Schlessinger. 1989. *Infrared Technology Fundamentals.* New York: Marcel Dekker. ISBN 0-8247-8134-1.

Spiro, I., ed. 1990. *Selected Papers on Radiometry.* Bellingham, MA: SPIE. ISBN 0-8194-03778-4.

Stover, J. 1990. *Optical Scattering.* New York: McGraw-Hill. ISBN 0-070-61814-3.

Tyson, R. 1991. *Principles of Adaptive Optics.* Orlando, FL: Academic Press. ISBN 0-127-05900-8.

Vincent, J. 1990. *Fundamentals of Infrared Detector Operation and Testing.* New York: John Wiley & Sons. ISBN 0-471-50272-3.

Waldman, G., and J. Wootton. 1992. *Electro-Optical Systems Performance Modeling.* Norwood, MA: Artech House. ISBN 0-890-006554-3.

Walker, G. 1989. *Miniature Refrigerators for Cryogenic Sensors and Cold Electronics.* Oxford, U.K.: Clarendon. ISBN 0-198-54815-X.

Weichel, H. 1988. *Laser System Design,* SPIE Lecture Notes. Bellingham, MA: SPIE.

Welford, W. 1991. *Useful Optics.* Chicago: University of Chicago Press. ISBN 0-226-89306-5.

Williams, C., and O. Becklund. 1972. *Optics: A Short Course for Engineers and Scientists.* New York: Wiley-Interscience. ISBN 0-471-94830-6.

Willardson, R., and A. Beer, eds. 1970. *Infrared Detectors.* New York: Academic Press.

Wolfe, W., ed. 1965. *Handbook of Military Infrared Technology.* Washington D.C.: Government Printing Office.

Wolfe, W., and G. Zissis, eds. 1978. *The Infrared Handbook.* Ann Arbor, MI: ERIM. ISBN 0-960-35901-X.

Wyatt, C. 1978. *Radiometric Calibration: Theory and Methods.* Orlando, FL: Harcourt Brace Jovanovich. ISBN 0-127-66150-6.

Wyatt, C. 1987. *Radiometric System Design.* New York: Macmillan. ISBN 0-029-48800-1.

Yakushenkov, Y. 1983. *Electro-Optical Devices.* Moscow: Mir.

Yoder, P. 1986. *Opto-Mechanical Systems Design.* New York: Marcel Dekker.

Index

About the Authors

JOHN LESTER MILLER earned a B.S. in Physics at the University of Southern California and an M.B.A. from Regis University. He has held positions as Program Manager, Functional Manager, Lead Engineer, and Electro-Optical Engineer with Lockheed Martin (Denver, Utica, and Orlando), Westinghouse (Orlando), University of Hawaii's NASA IRTF (Hilo), Rockwell International (Seal Beach), and Mt. Wilson and Palomar Observatories (Pasadena). He has published numerous papers and is author of *Principles of Infrared Technology*. Mr. Miller is known for creative solutions coupled to strict adherence to budgets and schedules, resulting in six corporate awards and recognition as an international authority. John's experience includes work on scientific instrumentation, IR surveillance systems, IR cameras, laser systems, fiber optics, IR search-and-track systems, environmental and weather monitoring sensors, long-life cryocoolers, advanced focal planes, and image processing. Mr. Miller, who now resides in a suburb of Portland, Oregon, is the Director of Advanced Technology for FLIR Systems Inc.

ED FRIEDMAN earned a B.S. in Physics at the University of Maryland in 1966 and a Ph.D. in cryogenic physics from Wayne State University in 1972. He started his career in the field of ocean optics and subsequently developed system concepts for remote sensing of the atmosphere and oceans. After completing studies related to the design of spacecraft and instruments for the measurement of the radiation balance of the Earth, he was appointed a Visiting Scientist in the climate program at the National Center for Atmospheric Research in Boulder, Colorado. On returning to industry, he concentrated on the military applications of visible and infrared sensors. He has been employed by The Mitre Corporation, Martin Marietta, and Ball Aerospace and Technologies Corporation. Most recently, he has been manager of a laboratory that performs performance assessments of complex optical and infrared sensor systems. Ed lives in the hills west of Denver.